INTRODUCTION TO THE GEOMETRY OF COMPLEX NUMBERS

INTRODUCTION TO THE GEOMETRY OF COMPLEX NUMBERS

BY

ROLAND DEAUX

Professor of Mathematics
La Faculté Polytechnique de Mons (Belgium)

Translated from the revised French edition
by

HOWARD EVES, Ph.D.

Professor of Mathematics
University of Maine

FREDERICK UNGAR PUBLISHING CO.
NEW YORK

Library of Congress Catalog Card Number 56-13198

Translated from the French
Introduction à la géométrie des nombres complexes
Revised and enlarged by the author

PRINTED IN BELGIUM

PREFACE TO THE AMERICAN EDITION

The presentation of this book to English-speaking students permits us to return to its reason for existence.

It is addressed to beginners, if we understand by this term all those who, whatever might be their mathematical knowledge beyond the usual studies, have not considered the use of complex numbers for establishing geometric properties of plane figures.

We declare, in fact, that a student accustomed to the classical methods of analytic geometry or of infinitesimal geometry is not, *ipso facto*, prepared to solve problems, even be they elementary, by appealing to complex numbers. He is even very often much embarrassed.

Now such obligations arise in the applied sciences, particularly in electrical studies, where the graphical interpretation of the calculation is a convenient picture of the circumstances of the phenomenon studied.

Finally, to assure the reader an easy and complete understanding, we have not hesitated to develop things from the beginning, by placing emphasis on the constructions connected with the algebraic operations.

In deference to the practical end endorsed, we have only considered constructions related to the concepts and methods of elementary geometry. One will find in other works (see, for example, the first cited book by Coolidge) constructions which, following the ideas of Klein, possess the character of invariance under the group of circular transformations.

The French edition has undergone a complete revision in the theory of unicursal bicircular quartics, in that of antigraphies, and in the proof of the theorem of Schick (art. 128).

In addition, the statements of 136 exercises have been inserted after the various sections. They will allow the reader to test his new knowledge and to interest himself at the same time with some aspects of the geometry of the triangle; a theory is well understood and will acquire power only after its successful application to the solution of proposed problems. At the start, some indications of the path to follow, but not always suppressing personal effort, are offered to assist the solver. One can find in the Belgian journal *Mathesis*, when such a reference is given, a complete solution by means of complex numbers and often accompanied by developments which enlarge the horizon of the problem.

The reader desirous of increasing his geometric knowledge in the domain to which the present work serves as an introduction will consult with profit, in addition to the books already cited, the remarkable work *Inversive Geometry* (G. Bell and Sons, Ltd., London, 1933) by Frank Morley and F. V. Morley. He will also find in *Advanced Plane Geometry* (North-Holland Publishing Company, Amsterdam, 1950), by C. Zwikker, a study of numerous curves by means of complex coordinates.

It is pleasant to express our gratitude to Professor Howard Eves of the University of Maine, who was so willing to take on the burdensome task of the translation and whose judicious remarks have improved the texture of several proofs, and our thanks to the Frederick Ungar Publishing Company of New York for the attentive care that it has brought to the presentation of this book.

Virelles-lez-Chimay (Belgium)
August, 1956

ROLAND DEAUX

FOREWORD

The present work briefly develops the lectures which we have given since 1930 to the engineering candidates who chose the section of electromechanics at the *Faculté polytechnique de Mons*.

A memoir by Steinmetz[1] emphasized the simplifying role that can be played by the geometric interpretation of complex numbers in the study of the characteristic diagrams of electrical machines. This idea has proved fruitful in its developments, and many articles published in such journals as the *Archiv für Elektrotechnik* can be comprehended only by the engineer acquainted with the meaning and the results of a plane analytic geometry adapted to representations in the Gauss plane. One there encounters, for example, unicursal bicircular quartics discussed by commencing with their parametric equation in complex coordinates; the procedure consisting of returning to classical geometry by the separation of the real and the imaginary will more often than not be devoid of interest, for it hides the clarity of certain interpretations which employ the relation between a complex number and the equivalent vectors capable of representing this number.

The study of such questions explains the appearance, particularly in Switzerland and Germany, of works in which geometric expositions accompany considerations on alternating currents.[2]

There does not exist a treatise of this sort in Belgium, and this lack will perhaps justify the book which we are presenting to the public.

Since, on the other hand, the geometrical results are independent of the technical reasons which called them forth, the service rendered to engineers will be little diminished if we stay in the purely mathematical domain. We most certainly do not wish to affirm by this that some technical illustrations of the theories which we

---

[1] *Die Anwendung komplexer Größen in der Elektrotechnik*, ETZ, 1893, pp. 597, 631, 641, 653.

[2] O. Bloch, *Die Ortskurven der graphischen Wechselstromtechnik*, Zürich, 1917.

G. Hauffe, *Ortskurven der Starkstromtechnik*, Berlin, 1932.

G. Oberdorfer, *Die Ortskurven der Wechselstromtechnik*, Munich, 1934.

develop, presented by an informed specialist, would be lacking in interest or timeliness; on the contrary, all our best wishes attend the efforts of the engineer who would contribute to the papers of Belgian scientific activity a work which certain foreign countries have judged useful to undertake.

An engineer looking for a geometric solution to a problem born in an electrotechnical laboratory will be able, in general, to limit himself to the first two chapters. It is interesting, however, to note that H. Pflieger-Haertel[1] arrived at the determination of the affix of the center of a circle, not by the method that we give for finding the foci of a conic (article 60), but by utilizing a property of Möbius transformations (article 127); and this shows that the engineer is able to take advantage of these theories too.

The table of contents states the subject matter of each of the articles, and therefore it appears unnecessary here to take up again the detailed sequence of material treated.

To the friends of geometry, we hope they find in the reading of these pages the pleasure that we have experienced in writing them. No serious difficulty needs to be surmounted. Variety is assured by the appeals that we make to algebra, to the classical notions of analytic geometry, to modern plane geometry, and to some results furnished by kinematics, and the third chapter revives in a slightly modified form the essentials of the projective geometry of real binary forms, thus putting in relief all the pertinency of the title given by Möbius to the first of his papers on the subject:[2]

Ueber eine Methode, um von Relationen, welche der Longimetrie angehören, zu entsprechenden Sätzen der Planimetrie zu gelangen (1852).

We would like to emphasize, moreover, the importance of those circular transformations of which the particular case of inversion is so often considered, but of which the direct involutoric case does not seem to have received the same favor, which it so well deserves.

The obligation of limiting ourselves has not permitted us to treat the antigraphy with the same amplification as the homography in the complex plane. Concerning these matters the reader will consult with profit the *Vorlesungen über projektive Geometrie* (Berlin, 1934) of C. Juel, the two books of J. L. Coolidge, *A treatise on the circle*

---

[1] *Zur Theorie der Kreisdiagramme: Archiv für Elektrotechnik,* vol. XII, 1923, pp. 486-493.

[2] *Werke*, 2, pp. 191-204.

and the sphere (Oxford, 1916), *The geometry of the complex domain* (Oxford, 1924), as well as the *Leçons de géométrie projective complexe* (Paris, 1931) of E. Cartan.

At a time when difficulties of all sorts create much anxiety for the most cautious editors, the firm of A. De Boeck has not, however, been afraid to undertake the publication of this work. For its noble courage, and for the attentive care that it has brought to the composition as well as to the presentation of this book, we express our thanks and our gratitude.

R. Deaux

Mons, October 23, 1945

TABLE OF CONTENTS

Chapter One

GEOMETRIC REPRESENTATION OF COMPLEX NUMBERS

Chapter Two

ELEMENTS OF ANALYTIC GEOMETRY IN COMPLEX NUMBERS

CHAPTER THREE

CIRCULAR TRANSFORMATIONS

CHAPTER ONE

GEOMETRIC REPRESENTATION OF COMPLEX NUMBERS

I. FUNDAMENTAL OPERATIONS

1. Complex coordinate. Consider the complex number

$$x + iy \qquad (x \text{ and } y \text{ real})$$

which we denote by z. Draw, in a plane, two perpendicular coordinate axes Ox, Oy. The point Z having for abscissa the real part x of the number z and for ordinate the coefficient y of i is called the *representative point*, or the *image*, of the number z. Conversely, each *real* point Z_1 of the plane is the image of a unique complex number z_1 equal to the abscissa of Z_1 increased by the product with i of the ordinate of this point. The number z_1 is called the *complex coordinate*, or the *affix*, of the point Z_1.

A plane in which each **real** point is considered as the image of a complex number is called the *Gauss plane*, the *Cauchy plane*, or the *plane of the complex variable.*

We shall denote a point of the Gauss plane by an upper case letter, and its affix by the corresponding lower,case letter.

Corollaries. 1° *The* Ox *axis is the locus of the images of the real numbers. The* Oy *axis is the locus of the images of the pure imaginary numbers.* This is why Ox and Oy are sometimes called the real axis and the imaginary axis of the Gauss plane.

2° *The number* $-z$ *is the affix of the symmetric of point* Z *with respect to the origin* O.

2. Conjugate coordinates. The complex number conjugate to

$$z = x + iy$$

will be designated by the notation

$$\bar{z} = x - iy$$

which is read, "$\bar{z}$ bar," and which will never appear as the written representation of a vector. A vector will always be indicated by the juxtaposition of the two upper case letters representing its origin and its extremity, surmounted by a bar, as $\overline{AB}$.

The image $\bar{Z}$ *of the number* $\bar{z}$ *is the point symmetric to* Z *with respect to the* Ox *axis.*

3. Exponential form. In the Gauss plane, we choose for the positive sense of rotation and of angles the sense of the smallest rotation about O which carries the Ox axis into the Oy axis. The algebraic value (xy) of any angle having Ox for initial side and Oy for terminal side is, then, to within an integral multiple of 2π,

$$(xy) = +\frac{\pi}{2}. \tag{1}$$

Let Z be the image of a non-zero complex number

$$z = x + iy. \tag{2}$$

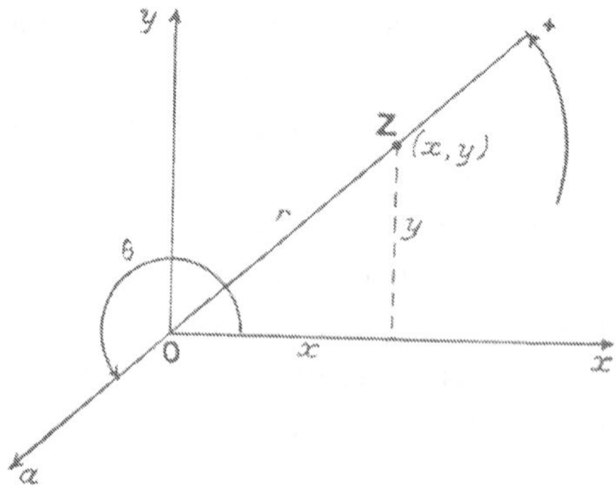

Fig. 1

Place an *arbitrary axis* a on the line OZ.

Denote by θ and r the algebraic values of any one of the angles (xa) and of the segment OZ. One recognizes in θ and r polar coordinates of Z for the pole O and the axis Ox. From the theorem on orthogonal projections we have

$$x = r\cos\theta$$

and, because of equation (1),

$$y = r\sin\theta.$$

Equation (2) can then be written as

$$z = r(\cos\theta + i\sin\theta)$$

or

$$\boxed{z = re^{i\theta},} \tag{3}$$

the exponential form of z.

When $z = 0$, we are to take r zero and θ arbitrary.

We have

$$\bar{z} = re^{-i\theta}.$$

4. Case where r is positive. When the positive sense of the a axis is that from O toward Z, the number r is positive and is the *modulus* of z; θ is then an *argument* of z. We write

$$r = |z| = |x + iy| = +\sqrt{x^2 + y^2},$$

the radical signifying that we extract the arithmetic square root of $x^2 + y^2$.

If, on the contrary, the a axis is such that r is negative, then equation (3) can be written as

$$z = (-r)(-1)\, e^{i\theta}$$

or, since $-1 = e^{i\pi}$, as

$$z = (-r)\, e^{i(\pi+\theta)}.$$

The modulus and an argument of z are then $-r$ and $\pi + \theta$.

5. Vector and complex number. The image Z of the number z is determined if we know the vector $\overline{OZ}$, the vectorial coordinate of Z for the pole O. We can then say that the number z is *represented by* this vector. The number and the vector have equal moduli, and we can conveniently speak of an argument of the number as an argument of the vector.

Nevertheless, we will never convey these facts by writing

$$\overline{OZ} = z$$

as is done by some authors, for such a use of the $=$ sign easily leads to contradictions when employed in connection with the product of two vectors (see article **10**) in the sense of classical vector analysis.

6. Addition. *If the* n *complex numbers*

$$z_k = x_k + iy_k \qquad (k = 1, 2, \dots, n)$$

have for images the n *points* Z_k, *their sum*

$$z = z_1 + z_2 + \dots + z_n \tag{4}$$

has for image the point Z *defined by the geometric equation*

$$\overline{OZ} = \overline{OZ}_1 + \overline{OZ}_2 + \dots + \overline{OZ}_n. \tag{5}$$

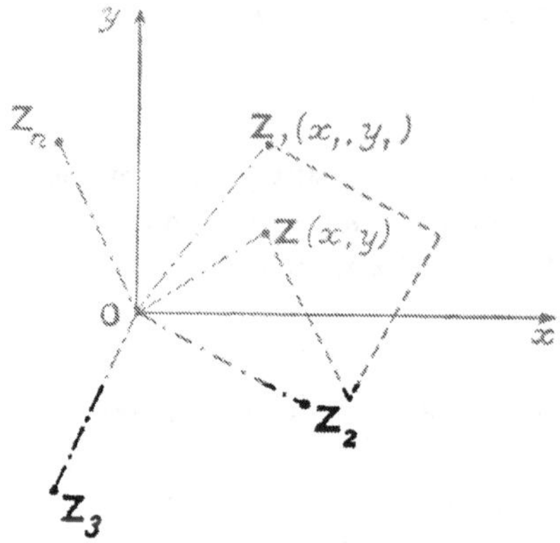

FIG. 2

Let x, y be the coordinates of the point Z constructed with the aid of equation (5). By algebraically projecting first on the Ox axis, then on the Oy axis, we obtain the two algebraic equations

$$x = \Sigma x_k, \qquad y = \Sigma y_k$$

and hence

$$x + iy = \Sigma (x_k + iy_k),$$

that is to say, equation (4).

7. Subtraction. *If the complex numbers* z_1, z_2 *are represented by the vectors* $\overline{OZ_1}$, $\overline{OZ_2}$, *the difference*

$$z = z_1 - z_2$$

is represented by the geometric difference

$$\overline{OZ} = \overline{OZ_1} - \overline{OZ_2}$$

of the corresponding vectors.

We have

$$z = z_1 + (-z_2) = z_1 + z_2'.$$

The point Z_2' is the symmetric of Z_2 with respect to O.

By virtue of article **6** we have

$$\overline{OZ} = \overline{OZ_1} + \overline{OZ_2'} = \overline{OZ_1} - \overline{OZ_2}. \quad (6)$$

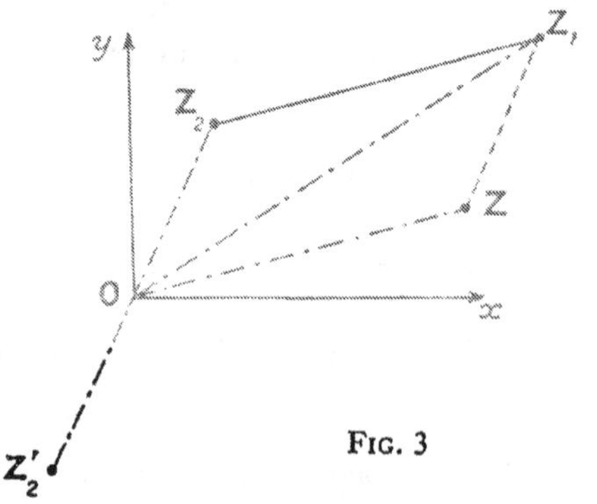

FIG. 3

Corollaries. 1° *Any vector of the Gauss plane represents the complex number equal to the complex coordinate of the extremity of the vector diminished by the complex coordinate of the origin of the vector.*

Equation (6) gives, in effect,

$$\overline{OZ} = \overline{Z_2 Z_1},$$

so that the vector $\overline{Z_2Z_1}$ represents, like $\overline{OZ}$, the difference $z_1 - z_2$.

2° *Any equality between two geometric polynomials whose terms are vectors of the Gauss plane is equivalent to an equality between two algebraic polynomials whose terms are the complex numbers represented by these vectors, and conversely.*

Thus, using the notation suggested in article 1, an equation such as

$$\overline{AB} = \overline{CD} + \overline{EF}$$

is equivalent to the algebraic equation

$$b - a = d - c + f - e.$$

8. Multiplication. *If the complex numbers z_1, z_2 are represented by the vectors $\overline{OZ_1}$, $\overline{OZ_2}$, the product*

$$z = z_1 z_2$$

is represented by the vector $\overline{OZ}$ which one obtains from $\overline{OZ_1}$, for example, as follows : 1° *rotate $\overline{OZ_1}$ about O through an angle equal to the argument of the other vector $\overline{OZ_2}$;* 2° *multiply the vector thus obtained by the modulus of vector $\overline{OZ_2}$.*

If r_1, r_2 and θ_1, θ_2 are the moduli and the arguments of z_1, z_2, we have, by (3) of article 3,

$$z_1 = r_1 e^{i\theta_1}, \qquad z_2 = r_2^{i\theta_2}.$$

Therefore

$$z = z_1 z_2 = r_1 r_2 e^{i(\theta_1+\theta_2)}.$$

The argument of z is then $\theta_1 + \theta_2$, while its modulus is

$$r_1 r_2 = | OZ_1 . OZ_2 |$$

which justifies the indicated construction.

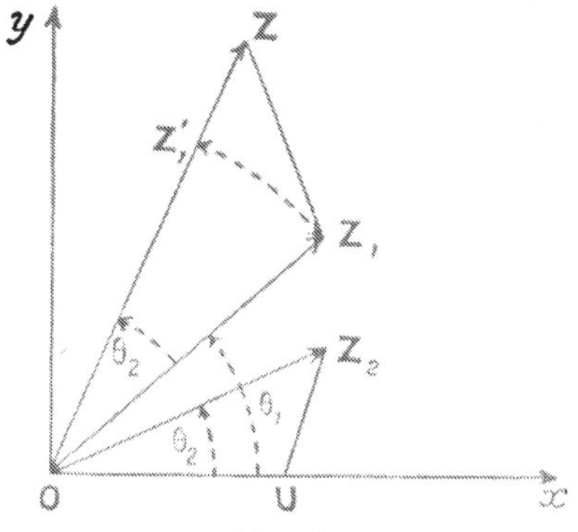

FIG. 4

We can realize the same end by taking on Ox the point U having abscissa + 1. The sought point Z is the third vertex of triangle OZ_1Z directly similar to triangle OUZ_2, for

$$(Ox, \overline{OZ}) = \theta_1 + \theta_2, \quad \left|\frac{OZ}{OZ_2}\right| = \frac{| OZ_1 |}{OU = 1}.$$

Particular cases. 1° *The number z_2 is real.* Its argument is 0 or π according as it is positive or negative.

We have

$$\overline{OZ} = z_2 \overline{OZ_1}$$

and the point Z is on the line OZ_1.

2° *The number* z_2 *is complex with unit modulus.* It then has the form

$$z_2 = e^{i\theta_2}, \qquad \theta_2 \neq k\pi, \qquad k \text{ an integer.}$$

The point Z is at Z_1' and is obtained from Z_1 by a *rotation* of angle θ_2 about O.

3° *The number* z_2 *is* i *or* — i. Since

$$i = e^{i\pi/2}, \qquad -i = e^{-i\pi/2},$$

to multiply a complex number z_1 *by* $\pm$ i *is to rotate its representative vector* $\overline{OZ_1}$ *about* O *through an angle of* $\pm \pi/2$.

9. Division. *If the complex numbers* z_1, z_2 *are represented by the vectors* $\overline{OZ_1}$, $\overline{OZ_2}$, *the quotient*

$$z = \frac{z_1}{z_2}$$

is represented by the vector $\overline{OZ}$ *which one obtains from the vector* $\overline{OZ_1}$ *as follows:* 1° *rotate* $\overline{OZ_1}$ *about* O *through an angle equal to the* **negative** *of the argument of vector* $\overline{OZ_2}$; 2° *divide the vector thus obtained by the modulus of vector* $\overline{OZ_2}$.

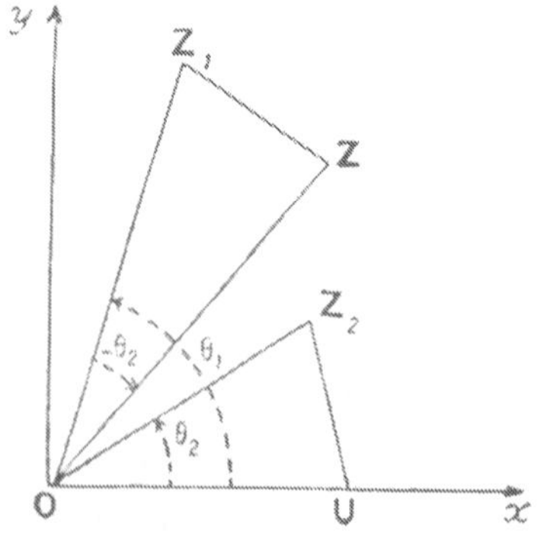

FIG. 5

Using the notation of article 8, the construction of Z follows from

$$z = \frac{r_1}{r_2} e^{i(\theta_1 - \theta_2)}.$$

The point Z is the third vertex of triangle OZ_1Z directly similar to triangle OZ_2U.

We thus treat division as the inverse operation of multiplication.

Particular case. *Construction of the point* Z *given by* $z = 1/z_2$. Since $z_1 = 1$, the point Z_1 is at U. The lines OZ_2, OZ are symmetric with respect to Ox and we have

$$|OZ| = \frac{1}{|OZ_2|} \quad \text{or} \quad |OZ_2 . OZ| = 1.$$

Then, if $\bar{Z}$ is the symmetric of Z with respect to Ox, the points Z_2, $\bar{Z}$ correspond to one another under the inversion having center O and power 1; that is to say, they are harmonic conjugates with respect to the extremities P, Q of the diameter of the circle having center O and radius OU.

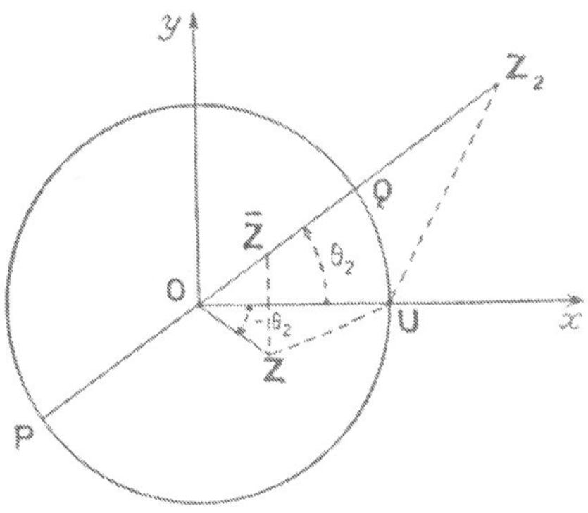

FIG. 6

Consequently, *the point* Z *is the symmetric with respect to* Ox *of the inverse of point* Z_2 *under the inversion having center* O *and power* 1.

Corollaries. 1° *If a point* A *has affix* a, *then its inverse under the inversion having center* O *and power* 1 *has affix* $1/\bar{a}$ *and not* $1/a$.

2° The construction of the point with affix

$$\frac{z_1}{z_2}$$

can be reduced to that of article 8 by writing

$$\frac{z_1}{z_2} = z_1 \times \frac{1}{z_2}.$$

10. Scalar product of two vectors. *If* z_1, z_2 *are the affixes of the points* Z_1, Z_2, *then*

$$\overline{OZ_1} \,.\, \overline{OZ_2} = \frac{1}{2}(z_1\bar{z}_2 + \bar{z}_1 z_2).$$

In fact, from the relations

$$z_1 = x_1 + iy_1, \qquad \bar{z}_1 = x_1 - iy_1$$

we find

$$x_1 = \frac{1}{2}(z_1 + \bar{z}_1), \quad y_1 = \frac{1}{2i}(z_1 - \bar{z}_1),$$

and, in the same way,

$$x_2 = \frac{1}{2}(z_2 + \bar{z}_2), \quad y_2 = \frac{1}{2i}(z_2 - \bar{z}_2).$$

It suffices to replace x_1, y_1, x_2, y_2 by these values in the classical expression

$$x_1 x_2 + y_1 y_2$$

for the product $\overline{OZ_1} \,.\, \overline{OZ_2}$ in order to obtain the announced result.

Corollary. *If* a *is the affix of the point* A, *and if* $\overline{AB}$, $\overline{CD}$ *are two vectors in the Gauss plane, then* (**7**, corollary 1°)

$$\overline{AB} \cdot \overline{CD} = \frac{1}{2}[(b-a)(\bar{d}-\bar{c}) + (\bar{b}-\bar{a})(d-c)].$$

11. Vector product of two vectors. *If* z_1, z_2 *are the affixes of the points* Z_1, Z_2, *the algebraic value of* $\overline{OZ_1} \times \overline{OZ_2}$ *on an axis* $O\zeta$ *such that the trihedral* $Oxy\zeta$ *is trirectangular and right-handed, is equal to*

$$\frac{i}{2}(z_1\bar{z}_2 - \bar{z}_1 z_2).$$

This algebraic value is, in fact,

$$x_1 y_2 - x_2 y_1$$

and it suffices to replace x_1, x_2, y_1, y_2 by the values given in article **10** in order to obtain the announced result.

Corollaries. 1° *The algebraic value considered is twice that of the area of triangle* OZ_1Z_2.

2° *If* a *is the affix of point* A, *and if* $\overline{AB}$, $\overline{CD}$ *are two vectors of the Gauss plane, we have* (**7**, corollary 1°) *for the algebraic value of* $\overline{AB} \times \overline{CD}$ *on an axis* $O\zeta$ *such that the trihedral* $Oxy\zeta$ *is trirectangular and right-handed*

$$\frac{i}{2}[(b-a)(\bar{d}-\bar{c}) - (\bar{b}-\bar{a})(d-c)].$$

12. Object of the present course. To each complex number z corresponds a point Z, and conversely. To each of the fundamental operations performed on complex numbers corresponds a geometric construction (**6-9**). Consequently, to each algebraic operation performed on complex numbers z_1, z_2, ..., z_n, and to each property of such an operation, corresponds a construction concerning the points Z_1, Z_2, ..., Z_n and a property of the figure obtained.

To interpret this passage from the algebraic manipulation of complex numbers to the geometric concept, or the reverse, is the aim of the present notes.

Exercises 1 through 11

1. Construct the sets of points having for affixes:

$1, i, -1, -i; 0, 1 + i, i\sqrt{2}, i - 1; 1 + i, i + 2, 2(1 + i), 1 + 2i; 1, -1, i\sqrt{3};$

$$1, \quad -1, \quad \frac{1 + i\sqrt{3}}{2}, \quad \frac{i\sqrt{3} - 1}{2}, \quad -\frac{1 + i\sqrt{3}}{2}, \quad \frac{1 - i\sqrt{3}}{2}.$$

2. Construct the images of the roots of the following equations:

$$z^2 - 1 = 0, \ z^2 + 1 = 0, \ z^3 - 1 = 0, \ z^3 + 1 = 0, \ z^4 - 1 = 0, \ z^4 + 1 = 0,$$
$$z^6 - 1 = 0, \quad z^6 + 1 = 0, \quad z^8 - 1 = 0, \quad z^8 + 1 = 0.$$

If n is a positive integer and a a number with modulus r and argument θ, what can be said about the figure formed by the images of the roots of the equations

$$z^n - a = 0, \qquad z^n + a = 0?$$

3. Which of the numbers represented by the following expressions are real and which are pure imaginary?

$$z + \bar{z}, \quad z - \bar{z}, \quad z\bar{z}, \quad z^2 - \bar{z}^2, \quad \frac{\left(\frac{1}{z} + \frac{1}{\bar{z}}\right)(z + \bar{z})}{z - \bar{z}},$$

$$z_1\bar{z}_2 - z_2\bar{z}_1, \quad z_1\bar{z}_2 + \bar{z}_1 z_2, \quad \frac{a\bar{b} + \bar{a}b}{a\bar{a} - 1},$$

$$\frac{a\bar{b} - \bar{a}b}{i(a\bar{a} + b\bar{b})}, \quad \frac{i(a\bar{b} + \bar{a}b)}{a\bar{b} - \bar{a}b}.$$

4. If a is the affix of a point A, construct the points with affixes:

$$-a, \quad \bar{a}, \quad \frac{1}{a}, \quad \frac{1}{\bar{a}}, \quad a + \bar{a}, \quad a - \bar{a}, \quad ia, \quad \frac{a}{\bar{a}}, \quad \frac{\bar{a}}{a}, \quad a + |a|,$$

$$a - |a|, \quad \frac{a}{|a|}, \quad \frac{|a|}{a}, \quad \frac{a - |a|}{a + |a|}, \quad \frac{a - |a|}{\bar{a} + |\bar{a}|}.$$

[Employ the exponential form in the last two cases.]

5. Distance between two points. The distance between two points A, B is

$$|\overline{AB}| = |b - a| = |a - b|.$$

Using the identity

$$x^2 + y^2 = (x + iy)(x - iy)$$

and the elements of analytic geometry, show that:

1° the square of the distance of a point Z from the origin O is $z\bar{z}$;

2° the square of the distance between points A and B is

$$(a - b)(\bar{a} - \bar{b}) \quad \text{or} \quad (b - a)(\bar{b} - \bar{a}).$$

6. Angles between two axes. Remembering that if two axes p, q have the senses of two vectors $\overline{AB}$, $\overline{CD}$ we have, where (p,q) denotes any one of the angles between these axes, the expression $|\overline{AB}|\,|\overline{CD}|\cos(p,q)$ for the scalar product $\overline{AB}\cdot\overline{CD}$, show that:

1° the angle between the vectors $\overline{Z_1Z_2}$, $\overline{Z_3Z_4}$ is given by

$$\cos(\overline{Z_1Z_2},\overline{Z_3Z_4}) = \frac{(z_2 - z_1)(\bar{z}_4 - \bar{z}_3) + (\bar{z}_2 - \bar{z}_1)(z_4 - z_3)}{\sqrt{(z_2 - z_1)(\bar{z}_2 - \bar{z}_1)}\ \sqrt{(z_4 - z_3)(\bar{z}_4 - \bar{z}_3)}},$$

where the radicals denote the arithmetic square roots;

2° the angle θ between vector $\overline{OZ}$ and the Ox axis is given by

$$\cos\theta = \frac{z + \bar{z}}{2\sqrt{z\bar{z}}}, \qquad \sin\theta = \frac{z - \bar{z}}{2i\sqrt{z\bar{z}}}, \qquad \tan\theta = \frac{\bar{z} - z}{\bar{z} + z},$$

expressions which are more easily obtained from $z = |z|\,e^{i\theta}$.

7. Equilateral triangle. A necessary and sufficient condition for three points A, B, C with affixes a, b, c to be vertices of an equilateral triangle is that

$$\frac{1}{b-c} + \frac{1}{c-a} + \frac{1}{a-b} = 0,$$

a relation which is equivalent to any one of the following:

$$a^2 + b^2 + c^2 = ab + bc + ca, \qquad (b - c)^2 + (c - a)^2 + (a - b)^2 = 0$$

$$(b - c)^2 = (c - a)(a - b), \qquad (c - a)^2 = (a - b)(b - c),$$

$$(a - b)^2 = (b - c)(c - a),$$

$$(b - c)(c - a) + (c - a)(a - b) + (a - b)(b - c) = 0,$$

$$\begin{vmatrix} a & b & 1 \\ b & c & 1 \\ c & a & 1 \end{vmatrix} = 0,$$

or, again, that $b - c$, $c - a$, $a - b$ be roots of an equation of the form

$$z^3 - k = 0.$$

[Set

$$b - c = \alpha, \qquad c - a = \beta, \qquad a - b = \gamma,$$

whence

$$\alpha + \beta + \gamma = 0.$$

It is necessary that

$$\alpha\bar{\alpha} = \beta\bar{\beta} = \gamma\bar{\gamma},$$

relations which, with

$$\bar{\alpha} + \bar{\beta} + \bar{\gamma} = 0,$$

give

$$\frac{1}{\alpha} + \frac{1}{\beta} + \frac{1}{\gamma} = 0.$$

Conversely, if

$$\alpha^2 = \beta\gamma,$$

whence

$$\bar{\alpha}^2 = \beta\bar{\gamma}, \qquad (\alpha\bar{\alpha})^2 = (\beta\bar{\beta})(\gamma\bar{\gamma}), \qquad (\alpha\bar{\alpha})^3 = (\alpha\bar{\alpha})(\beta\bar{\beta})(\gamma\bar{\gamma}),$$

then

$$\alpha\bar{\alpha} = \beta\bar{\beta} = \gamma\bar{\gamma}.$$

See article **84** for another demonstration.]

8. With the aid of exercise 7, show that the images of the roots of the cubic equation

$$z^3 + 3a_1z^2 + 3a_2z + a_3 = 0$$

form an equilateral triangle if $a_1^2 = a_2$. Deduce that the origin and the images of the roots of the equation

$$z^2 + pz + q = 0$$

form an equilateral triangle if $p^2 = 3q$.

9. The lines OZ_1, OZ_2 are perpendicular or parallel according as

$$z_1\bar{z}_2 + \bar{z}_1z_2 = 0 \qquad \text{or} \qquad z_1\bar{z}_2 = \bar{z}_1z_2.$$

The lines Z_1Z_2, Z_3Z_4 are perpendicular or parallel according as

$$(z_1 - z_2)(\bar{z}_3 - \bar{z}_4) + (\bar{z}_1 - \bar{z}_2)(z_3 - z_4) = 0$$

or

$$(z_1 - z_2)(\bar{z}_3 - \bar{z}_4) = (\bar{z}_1 - \bar{z}_2)(z_3 - z_4).$$

[Use articles **10** and **11**.]

10. If $ab = cd$, we have

$$|OA||OB| = |OC||OD|$$

and the angles $(\overline{OA},\overline{OB})$, $(\overline{OC},\overline{OD})$ have the same bisectors. [Use moduli and arguments.]

If $ab = c^2$, OC is the interior bisector of angle $(\overline{OA},\overline{OB})$. What can be said if $ab = -c^2$?

11. If a, b, c are numbers of modulus 1 and if we set

$$s_1 = a + b + c, \qquad s_2 = ab + bc + ca, \qquad s_3 = abc,$$

show, by employing

$$a\bar{a} = b\bar{b} = c\bar{c} = 1,$$

that

$$\bar{s}_1 = \frac{s_2}{s_3}, \quad \bar{s}_2 = \frac{s_1}{s_3}, \quad \bar{s}_3 = \frac{1}{s_3}, \quad s_1 = \frac{\bar{s}_2}{\bar{s}_3}, \quad s_2 = \frac{\bar{s}_1}{\bar{s}_3}, \quad s_1\bar{s}_1 = s_2\bar{s}_2, \quad \left|\frac{s_2}{s_1}\right| = 1,$$

and that triangle ABC is equilateral if $s_1^2 = 3s_2$.

II. FUNDAMENTAL TRANSFORMATIONS

13. Transformation. Any process associating with each point Z of the Gauss plane at least one point Z′ of this plane constitutes a transformation of the plane into itself, and may be designated by a letter, say ω.

We shall consider only transformations in which

1° to each point Z there corresponds a single point Z′;

2° each point Z′ is the correspondent of a single point Z.

Such a transformation ω is said to be *one-to-one;* the point Z′ is the *correspondent*, or the *homologue*, of Z.

The process which, starting with Z′, yields the point Z, is called the *inverse transformation* of ω and is denoted by ω^{-1}.

The *equation of a transformation* ω is the relation between the affix z of an arbitrary point Z of the plane and the affix z' of the point Z′ corresponding to Z.

14. Translation. Let A be a fixed point and Z an arbitrary point of the plane and let a and z be their affixes.

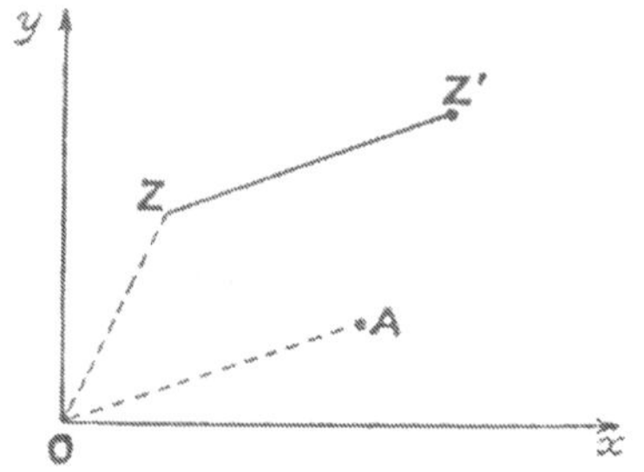

FIG. 7

The point Z′ such that

$$\overline{ZZ'} = \overline{OA}$$

is called the homologue of Z in the *translation of vector* $\overline{OA}$.

Since we have

$$\overline{OZ'} = \overline{OZ} + \overline{OA},$$

the *equation of the translation* is (7, corollary 2°)

$$\boxed{z' = z + a.}$$

When a is real but not zero, the translation is parallel to Ox. If $a = 0$, the translation reduces to the *identity transformation* with equation

$$z' = z$$

which transforms each point of the plane into itself.

15. Rotation. Let A be a fixed point of affix a, and let α be a given real number, positive, zero, or negative. In a rotation about A of an angle with algebraic value α, each point Z of the plane takes a position Z′.

The vectors $\overline{AZ'}$, $\overline{AZ}$ represent the complex numbers

$$z' - a, \qquad z - a$$

(7, corollary 1°), and since $\overline{AZ'}$ is obtained from $\overline{AZ}$ by the indicated rotation, we have (8, particular case 2°)

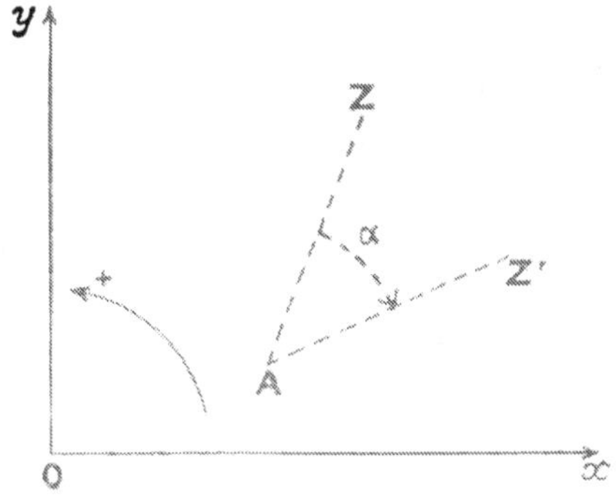

FIG. 8

$$z' - a = (z - a)\, e^{i\alpha}.$$

The equation of the rotation of angle α *in algebraic value, about the point of affix* a, *is then*

$$\boxed{z' = ze^{i\alpha} + a\,(1 - e^{i\alpha}).}$$

Corollary. The *symmetry with respect to the point* A is nothing but a rotation of angle $\alpha = \pi$ (or $-\pi$) about A and, since

$$e^{i\pi} = \cos \pi + i \sin \pi = -1,$$

its equation is

$$z' = -z + 2a.$$

16. Homothety. Let us be given a fixed point A of affix a and a real non-zero number k, positive or negative. If we arbitrarily place an axis on the line joining A to any point Z whatever of the plane, the point Z′ of this axis such that we have the algebraic relation

$$\frac{AZ'}{AZ} = k$$

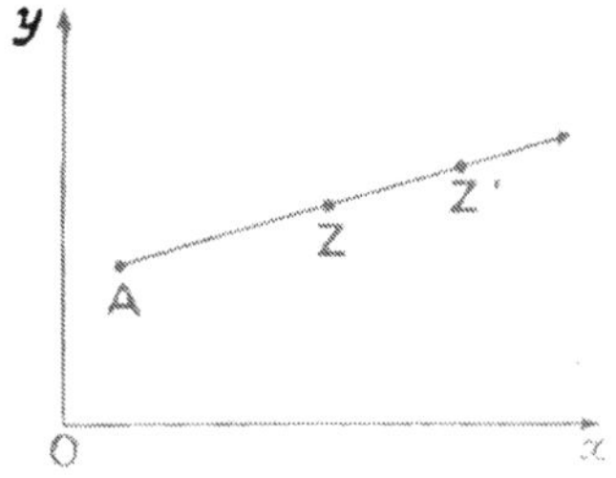

FIG. 9

is the homologue of Z in the homothety (A, k) of *center* A and *coefficient*, or *ratio*, k.[1]

Since

$$\overline{AZ'} = k\,\overline{AZ},$$

we also have (7)

$$z' - a = k\,(z - a)$$

[1] See R. DEAUX, *Compléments de géométrie plane*, article 116 (A. De Boeck, Brussels, Belgium, 1955).

and the *equation of the homothety* is

$$\boxed{z' = kz + a\,(1 - k).}$$

Remark. The values 1, — 1 for k give the identity transformation (14) and the symmetry with respect to point A (15).

17. Relation among three points. *Being given three points* A, B, C *with affixes* a, b, c, *if* AB, AC *are the algebraic values of segments calculated on axes* a_1, a_2 *arbitrarily placed on the lines* AB, AC, *we have*

$$c - a = (b - a)\, e^{i(a_1 a_2)} \cdot \frac{\mathrm{AC}}{\mathrm{AB}},$$

where $(a_1 a_2)$ *is the algebraic value of any one of the angles having* a_1 *for initial side and* a_2 *for terminal side.*

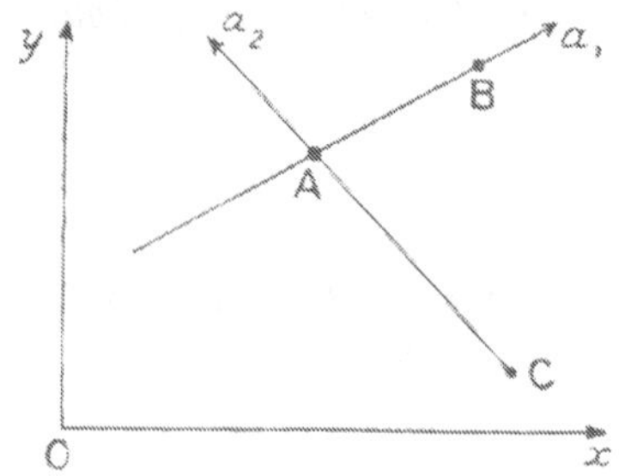

FIG. 10

We have the equations (3, 7 corollary 1°)

$$c - a = \mathrm{AC}\, e^{i(xa_2)}$$

$$b - a = \mathrm{AB}\, e^{i(xa_1)}$$

which give

$$\frac{c - a}{b - a} = \frac{\mathrm{AC}}{\mathrm{AB}}\, e^{i\,[(xa_2) - (xa_1)]}.$$

The angle relation of Möbius[1]

$$(a_1 a_2) = (xa_2) - (xa_1)$$

yields the announced equation.

18. Symmetry with respect to a line. *The line being given by two of its points* A, B, let Z′ be the symmetric with respect to this line of any point Z of the plane, let d_1 and d_2 be axes arbitrarily placed on the lines AZ, BZ, and let d_1', d_2' be axes which are the symmetrics of d_1, d_2 with respect to the line AB. We have (17)

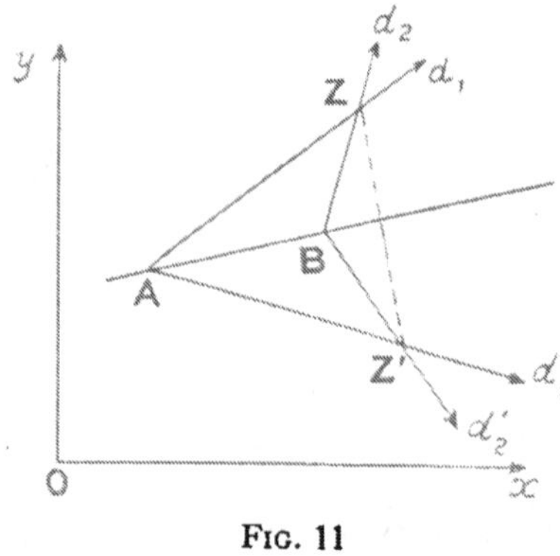

FIG. 11

$$a - z = (b - z)\, e^{i(d_2 d_1)} \cdot \frac{\mathrm{ZA}}{\mathrm{ZB}}, \quad (1)$$

[1] See R. DEAUX, *Compléments de géométrie plane*, article 84.

$$a - z' = (b - z')\, e^{i\,(d_2' d_1')} \cdot \frac{\mathrm{Z'A}}{\mathrm{Z'B}}, \tag{2}$$

$$\mathrm{ZA} = \mathrm{Z'A}, \quad \mathrm{ZB} = \mathrm{Z'B}, \qquad (d_2'\, d_1') = -(d_2\, d_1). \tag{3}$$

In order to eliminate from equations (1) and (2) the ratio ZA/ZB and the angle $(d_2 d_1)$, which depend on the point Z considered, let us substitute for the first equation the result obtained by replacing each number in the equation by its conjugate, that is to say, by

$$\bar{a} - \bar{z} = (\bar{b} - \bar{z})\, e^{-i\,(d_2\, d_1)} \cdot \frac{\mathrm{ZA}}{\mathrm{ZB}}. \tag{4}$$

If we divide equation (2) by equation (4), member by member, and take note of (3), we obtain the *equation of the symmetry* in the form

$$\frac{a - z'}{\bar{a} - \bar{z}} = \frac{b - z'}{\bar{b} - \bar{z}}$$

or

$$\boxed{\frac{a - z'}{b - z'} = \frac{\bar{a} - \bar{z}}{\bar{b} - \bar{z}}} \tag{5}$$

or, again,

$$z' = \frac{a - b}{\bar{a} - \bar{b}}\, \bar{z} - \frac{a\bar{b} - \bar{a}b}{\bar{a} - \bar{b}}. \tag{6}$$

Remark. When the line AB is the Ox-axis, the numbers a and b are real, and we have

$$\bar{a} = a, \quad \bar{b} = b$$

and equation (6) becomes

$$z' = \bar{z},$$

a result which follows immediately from article **2**.

19. Inversion. Let p be the power, positive or negative, of an inversion of center M with affix m; let d be an axis arbitrarily placed on the line joining M to any point Z of the plane; let Z′ be the inverse of Z. We have

$$\mathrm{MZ} \cdot \mathrm{MZ'} = p \tag{7}$$

and (3)

$$z - m = \mathrm{MZ} \cdot e^{i(xd)} \tag{8}$$

$$z' - m = \mathrm{MZ'} \cdot e^{i(xd)}. \tag{9}$$

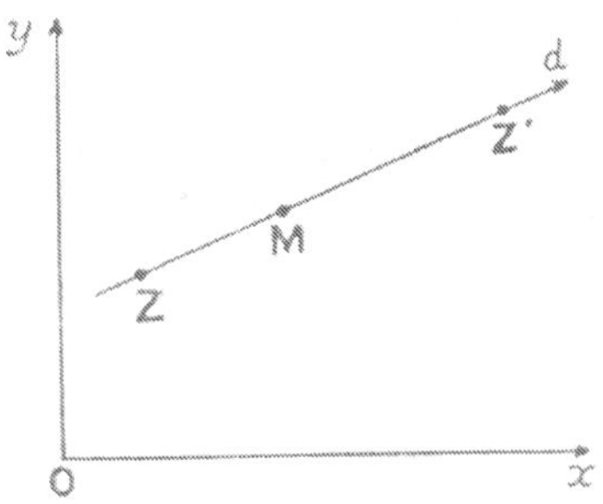

FIG. 12

In order to introduce the expression (7) for p, and at the same time to eliminate (xd), which depends on the point Z considered, let us substitute for equation (8) the result obtained by replacing each number in the equation by its conjugate, that is to say, by

$$\bar{z} - \bar{m} = \text{MZ}\, e^{-i(xd)}. \tag{10}$$

If we multiply equation (9) by equation (10), member by member, and take note of (7), we get, for the *equation of the inversion* (M, p) *of center* M *and power* p,

$$\boxed{(z' - m)(\bar{z} - \bar{m}) = p}$$

or

$$z' = \frac{m\bar{z} + p - m\bar{m}}{\bar{z} - \bar{m}}.$$

Remark. When M is at O and $p = 1$, the equation of the inversion is

$$z' = \frac{1}{\bar{z}},$$

which agrees with a corollary of article **9**.

20. Point at infinity in the Gauss plane. The projective plane (which can be studied in analytic geometry with the aid of homogeneous coordinates) contains an infinitude of points at infinity constituting a range of points lying on the line at infinity in this plane.

The Gauss plane (which is studied with the aid of the complex coordinate z) *contains*, on the contrary, *only a single point at infinity*, that which corresponds to z infinite.

By virtue of the equation (**19**)

$$z' = \frac{1}{\bar{z}},$$

in an inversion, the point at infinity in the Gauss plane is the inverse of the center of inversion.

21. Product of one-to-one transformations. If a transformation ω_1 associates with each point Z a point Z_1, we express this fact by writing

$$Z_1 = \omega_1 [Z]. \tag{11}$$

Consider a transformation ω_2 which, operating on each point Z_1, transforms this into a point Z_2; we will have

$$Z_2 = \omega_2 [Z_1]$$

or, taking note of equation (11),

$$Z_2 = \omega_2 \{\omega_1 [Z]\}$$

which we agree to write as

$$Z_2 = \omega_2 \omega_1 [Z]. \qquad (12)$$

The transformation ω permitting us to pass directly from the points Z to the points Z_2 is called the *product of the transformations* ω_1, ω_2 *taken in this order.*

Equation (12) and

$$Z_2 = \omega (Z)$$

lead us to write, by convention,

$$\omega = \omega_2 \omega_1. \qquad (13)$$

In the symbolic product $\omega_2\omega_1$, *the second factor* ω_1 *represents the first transformation to be performed.*

These considerations can be extended to any number of transformations. Furthermore, from an equation such as (13) we can form equations

$$\omega\omega_3 = \omega_2 \omega_1 \omega_3, \qquad \omega_3 \omega = \omega_3 \omega_2 \omega_1$$

obtained by multiplying the two members of (13), either on the right or on the left, by the same transformation ω_3; this follows from the definition of a product of transformations.

Examples. 1° Product of two translations. Let a_1, a_2 be complex numbers represented by the vectors $\overline{OA_1}$, $\overline{OA_2}$ which define the translations ω_1, ω_2 (14). If Z_1 is the correspondent of Z under ω_1, while Z_2 is that of Z_1 under ω_2, we have

$$z_1 = z + a_1, \qquad (14)$$

$$z_2 = z_1 + a_2. \qquad (15)$$

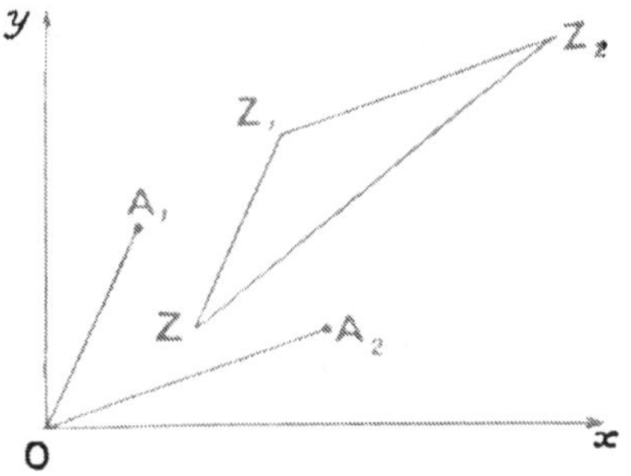

FIG. 13

The equation of the transformation $\omega_2\omega_1$ permitting the direct passage from Z to Z_2 is obtained by eliminating z_1 from equations (14) and (15), and is

$$z_2 = z + a_1 + a_2. \qquad (16)$$

This proves that the product $\omega_2\omega_1$ is a translation of vector $\overline{OA_1} + \overline{OA_2}$. This also follows because we have

$$\overline{ZZ_2} = \overline{ZZ_1} + \overline{Z_1Z_2} = \overline{OA_1} + \overline{OA_2}.$$

2° **Product of two rotations.** Let a_1, a_2 be the affixes of the centers A_1, A_2 of rotations of angles having algebraic values α_1, α_2. If Z_1 is the correspondent of Z in the first rotation ω_1, while Z_2 is that of Z_1 in the second rotation ω_2, we have

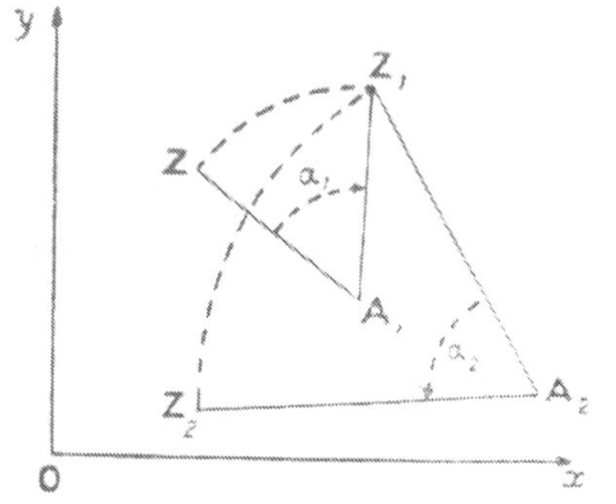

Fig. 14

(15)
$$z_1 - a_1 = (z - a_1)\, e^{i\,\alpha_1},$$
$$z_2 - a_2 = (z_1 - a_2)\, e^{i\,\alpha_2}.$$

The elimination of z_1 gives the equation

$$z_2 = ze^{i(\alpha_1+\alpha_2)} + a_1\,(1 - e^{i\,\alpha_1})\, e^{i\,\alpha_2} + a_2\,(1 - e^{i\,\alpha_2}) \qquad (17)$$

of the transformation $\omega_2\omega_1$ which binds Z_2 to Z, and which will be examined later (86).

3° *The product of a transformation* ω *and its inverse* ω^{-1} *is the identity transformation, and is represented by* 1. Thus we have

$$\omega^{-1}\omega = 1, \text{ and } \omega\omega^{-1} = 1.$$

22. Permutable transformations. Two transformations ω_1, ω_2 are said to be permutable if the correspondent of each point Z in the transformation $\omega_2\omega_1$ coincides with the correspondent of Z in the transformation $\omega_1\omega_2$. We write

$$\omega_2\omega_1 = \omega_1\omega_2.$$

Examples. 1° *Any two translations whatever are permutable.*

The equation of the product $\omega_1\omega_2$ of the translations considered in the first example of article **21** being

$$z' = z + a_2 + a_1,$$

proves, if we compare with equation (16), that $Z' = Z_2$, which is what we wished to establish.

2° *Two rotations are permutable only if they have the same center.*

The equation of the product $\omega_1\omega_2$ of the rotations considered in the second example of article **21** being

$$z' = ze^{i(\alpha_2+\alpha_1)} + a_2\,(1 - e^{i\,\alpha_2})\, e^{i\,\alpha_1} + a_1\,(1 - e^{i\,\alpha_1}),$$

proves, if we compare with equation (17), that the points Z', Z_2 coincide only if we have

$$a_1\,(1 - e^{i\,\alpha_1})\, e^{i\,\alpha_2} + a_2\,(1 - e^{i\,\alpha_2}) = a_2\,(1 - e^{i\,\alpha_2})\, e^{i\,\alpha_1} + a_1\,(1 - e^{i\,\alpha_1})$$

or

$$(a_1 - a_2)(1 - e^{i\alpha_1})(1 - e^{i\alpha_2}) = 0.$$

Since α_1, α_2 are supposed not to be integral multiples of 2π, the exponentials $e^{i\alpha_1}$, $e^{i\alpha_2}$ are different from 1, whence $a_1 = a_2$ and the centers A_1, A_2 of the rotations coincide.

23. Involutoric transformation. A transformation ω is said to be involutoric *if the correspondent of an arbitrary point Z being Z′, the correspondent of Z′ is Z.*

The product of the transformation ω and the transformation ω is then the identity transformation, and we have

$$\omega\omega = 1 \quad \text{or} \quad \omega^2 = 1. \tag{18}$$

We therefore say that *a transformation is involutoric if its square is the identity transformation.*

If we multiply the two members of the symbolic equation (18) by ω^{-1}, we have (21)

$$\omega\omega\omega^{-1} = \omega^{-1} \quad \text{or} \quad \omega = \omega^{-1}.$$

Consequently, *a transformation is involutoric if it is identical with its inverse.*

A symmetry with respect to a point, a symmetry with respect to a line, and an inversion are involutoric transformations.

24. Changing coordinate axes. Let $Ax'y'$ be a new rectangular system such that $(x'y') = (xy)$ and defined by the affix $a = a_1 + ia_2$ of A and by the algebraic value α of angle (xx').

The object is to find the relation between the affixes

$$z = x + iy, \quad z' = x' + iy'$$

of an arbitrary point $Z = Z'$ with respect to the two systems Oxy, $Ax'y'$.

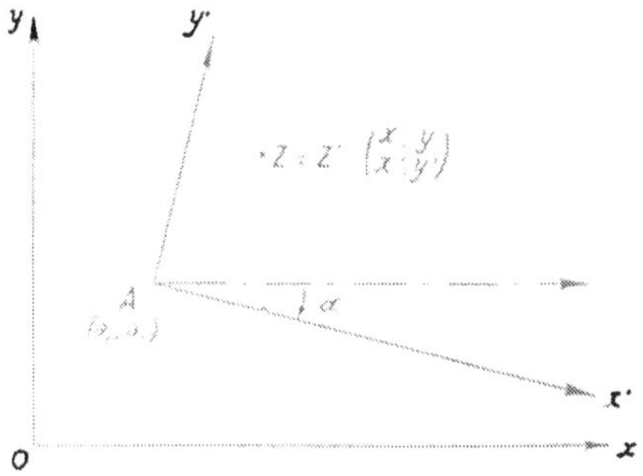

FIG. 15

We know from analytic geometry that

$$x = a_1 + x' \cos\alpha - y' \sin\alpha,$$
$$y = a_2 + x' \sin\alpha + y' \cos\alpha.$$

From this we obtain

$$x + iy = a_1 + ia_2 + x' (\cos \alpha + i \sin \alpha) + iy' (\cos \alpha + i \sin \alpha),$$

$$\boxed{z = a + z'e^{i\alpha}.}$$

Exercises 12 through 16

12. We are given the affixes a, b of the consecutive vertices A, B of two squares ABCD, ABC'D' and we know that the vertices A, B, C, D succeed each other in the positive sense of rotation. Find the affixes of the points C, D, C', D' and of the centers M, M' of the squares. [Employ rotations, translations, and homotheties,

$$c = b(1 + i) - ia, \qquad d = a(1 - i) + ib, \qquad c' = ia + b(1 - i),$$

$$d' = a(1 + i) - ib, \qquad m = \frac{b(1 + i)}{2} + \frac{a(1 - i)}{2}, \qquad m' = \frac{a(1 + i)}{2} + \frac{b(1 - i)}{2}.]$$

13. If a, b are the affixes of the vertices A, B of equilateral triangles ABC, ABC', find the affixes of C, C', knowing that angle $(\overline{CA,CB})$ is $-\pi/3$.

Show that

$$c + c' = a + b, \qquad cc' = a^2 + b^2 - ab,$$

and also obtain these results from a relation of exercise 7.

14. The product of two inversions with the same center is a homothety. Show that the inversions are permutable only if their powers are opposite and that the homothety is then a symmetry.

15. Two inversions having distinct centers are permutable only if the square of the distance between the centers is the sum of the powers of inversion, or, in other words, only if the director circles or circles of double points are orthogonal. Their product is then an involutoric transformation (a Möbius involution, see article **99**). [For symmetry in calculation, take Ox on the line of centers, ...]

16. An inversion is permutable with a symmetry with respect to a line only if its center is on the line. The product is then an involution. [Take the Ox axis on the line.]

III. *ANHARMONIC RATIO*

25. Definition and interpretation. The anharmonic ratio (A.R.) of four distinct points Z_1, Z_2, Z_3, Z_4 of the Gauss plane, taken in this order, is by definition that of their affixes z_1, z_2, z_3, z_4, and is denoted by

$$(Z_1\,Z_2\,Z_3\,Z_4) \quad \text{or} \quad (z_1\,z_2\,z_3\,z_4)$$

and has for value

$$\frac{z_1 - z_3}{z_2 - z_3} : \frac{z_1 - z_4}{z_2 - z_4}.$$

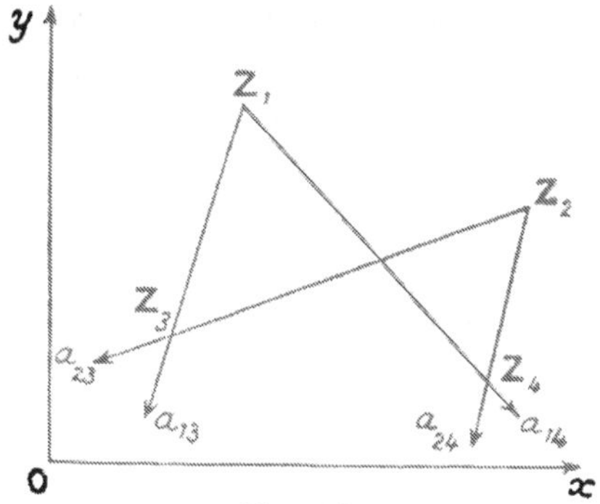

FIG. 16

In order *to express the* A.R. *as a function of geometric elements*, place arbitrary axes a_{13}, a_{14}, a_{23}, a_{24} on the lines Z_1Z_3, Z_1Z_4, Z_2Z_3, Z_2Z_4, which may or may not be distinct, and designate by Z_1Z_3, for example, the algebraic value calculated on a_{13} of the segment having Z_1 for initial point and Z_3 for terminal point. Assuming $(xy) = +\pi/2$, we have (17)

$$\frac{z_1 - z_3}{z_2 - z_3} = \frac{Z_3Z_1}{Z_3Z_2} e^{i(a_{23}a_{13})} \quad ; \quad \frac{z_1 - z_4}{z_2 - z_4} = \frac{Z_4Z_1}{Z_4Z_2} e^{i(a_{24}a_{14})},$$

$$\boxed{(z_1z_2z_3z_4) = (Z_1Z_2Z_3Z_4) = \left(\frac{Z_1Z_3}{Z_2Z_3} : \frac{Z_1Z_4}{Z_2Z_4}\right) e^{i[(a_{23}a_{13}) - (a_{24}a_{14})]}.}$$

If we should choose a_{13}, a_{14}, a_{23}, a_{24} so that

$$\frac{Z_1Z_3}{Z_2Z_3} : \frac{Z_1Z_4}{Z_2Z_4}$$

is positive, then this number is the modulus of the A.R. and an argument is $(a_{23}a_{13}) - (a_{24}a_{14})$.

This is certainly the case if we take

$$(z_1z_2z_3z_4) = \left(\frac{Z_1Z_3}{Z_2Z_3} : \frac{Z_1Z_4}{Z_2Z_4}\right) e^{i[(\overline{Z_2Z_3}, \overline{Z_1Z_3}) - (\overline{Z_2Z_4}, \overline{Z_1Z_4})]}.$$

26. Properties. The classic properties of the A.R. of four real or complex numbers permit us to state the following properties of the A.R. of four points of the Gauss plane.

1° *An* A.R. *of four points does not alter in value if we exchange two points and at the same time the other two points; it takes the reciprocal value if we exchange the first two points or the last two points; it takes the value complementary to unity if we exchange the two mean points or the two extreme points.*

2° *With four points we can form* 24 A.R.'*s, presenting at most* 6 *values, and* 3 *of these values are the reciprocals of the other* 3.

$$(Z_1Z_2Z_3Z_4) = (Z_2Z_1Z_4Z_3) = (Z_3Z_4Z_1Z_2) = (Z_4Z_3Z_2Z_1) = \lambda,$$

$$(Z_1Z_2Z_4Z_3) = (Z_2Z_1Z_3Z_4) = (Z_4Z_3Z_1Z_2) = (Z_3Z_4Z_2Z_1) = \frac{1}{\lambda},$$

$$(Z_1Z_3Z_2Z_4) = (Z_3Z_1Z_4Z_2) = (Z_2Z_4Z_1Z_3) = (Z_4Z_2Z_3Z_1) = 1-\lambda,$$

$$(Z_1Z_3Z_4Z_2) = (Z_3Z_1Z_2Z_4) = (Z_4Z_2Z_1Z_3) = (Z_2Z_4Z_3Z_1) = \frac{1}{1-\lambda},$$

$$(Z_1Z_4Z_2Z_3) = (Z_4Z_1Z_3Z_2) = (Z_2Z_3Z_1Z_4) = (Z_3Z_2Z_4Z_1) = \frac{\lambda-1}{\lambda},$$

$$(Z_1Z_4Z_3Z_2) = (Z_4Z_1Z_2Z_3) = (Z_3Z_2Z_1Z_4) = (Z_2Z_3Z_4Z_1) = \frac{\lambda}{\lambda-1}.$$

3° The A.R.'s

$$(Z_1Z_2Z_3Z_4) = \lambda, \quad (Z_1Z_3Z_4Z_2) = \frac{1}{1-\lambda}, \quad (Z_1Z_4Z_2Z_3) = \frac{\lambda-1}{\lambda}$$

obtained by keeping the first point fixed and by circularly permuting the other three are *three principal* A.R.'*s*.

4° If the *four points are distinct, their* A.R.'*s are different from* 1, 0, ∞.

5° We have

$$(Z_1Z_2Z_3Z_4) = \frac{\dfrac{1}{z_1-z_2} - \dfrac{1}{z_1-z_4}}{\dfrac{1}{z_1-z_2} - \dfrac{1}{z_1-z_3}},$$

a relation *which displays an* A.R. *as a function of the differences between one affix and each of the other three.*

27. Case where a point is at infinity. We denote by ∞ both the point at infinity in the Gauss plane and its affix.

By definition we have

$$(Z_1Z_2Z_3\infty) = \lim_{z_4\to\infty} (z_1z_2z_3z_4) = \lim_{z_4\to\infty} \left(\frac{z_1 - z_3}{z_2 - z_3} : \frac{z_1 - z_4}{z_2 - z_4}\right) = \frac{z_1 - z_3}{z_2 - z_3}.$$

Hence, *in order to develop an* A.R. *in which the point at infinity is not in the fourth place, we begin by bringing the point into this place* (**26**, 1°). Thus

$$(Z_1\infty\, Z_3Z_4) = (Z_3Z_4Z_1\infty) = \frac{z_3 - z_1}{z_4 - z_1}.$$

Corollary. Each real or imaginary number z is the anharmonic ratio determined by the point Z, the point U on the Ox-axis having abscissa 1, the origin O, and the point at infinity, for

$$(\mathrm{ZUO}\infty) = (z10\infty) = z.$$

28. Real anharmonic ratio. *In order that the* A.R. *of four points* Z_1, Z_2, Z_3, Z_4 *of the complex plane be real, it is necessary and sufficient that these points belong to a common line or to a common circle. This* A.R. *is then the same as that considered in elementary geometry.*

With the notation of article **25** we have

$$(Z_1Z_2Z_3Z_4) = \left(\frac{Z_1Z_3}{Z_2Z_3} : \frac{Z_1Z_4}{Z_2Z_4}\right) e^{i[(a_{23}a_{13}) - (a_{24}a_{14})]} \tag{1}$$

and for this A.R. to be real it is necessary and sufficient that for some integer n we have

$$(a_{23}\, a_{13}) - (a_{24}\, a_{14}) = n\pi. \tag{2}$$

Case 1 — *where the points* Z_1, Z_2, Z_3 *are collinear.* For some integer n_1 we have

$$(a_{23}\, a_{13}) = n_1\pi$$

and consequently

$$(a_{24}\, a_{14}) = (n_1 - n)\,\pi$$

which proves that point Z_4 is on the line Z_1Z_2.

Choose the axes a_{13}, a_{14}, a_{23}, a_{24} so that they form a single axis. Equation (1) then becomes

$$(Z_1Z_2Z_3Z_4) = \frac{Z_1Z_3}{Z_2Z_3} : \frac{Z_1Z_4}{Z_2Z_4}$$

and is the same as the definition given in elementary geometry for the A.R. of four collinear points.

Case 2—where the points Z_1, Z_2, Z_3 *are not collinear.* The points lie on the circle γ which they determine. Orient a_{13}, a_{23}, a_{14} from Z_3 toward Z_1, from Z_3 toward Z_2, and from Z_4 toward Z_1; then choose the positive sense of a_{24} so that in equation (2), which holds by hypothesis, the integer n is even. The equation then becomes

$$(a_{24}\, a_{14}) = (a_{23}\, a_{13}) + 2n_2\, \pi, \quad (n_2 \text{ an integer})$$

and proves that the point Z_4 is on the circle γ.

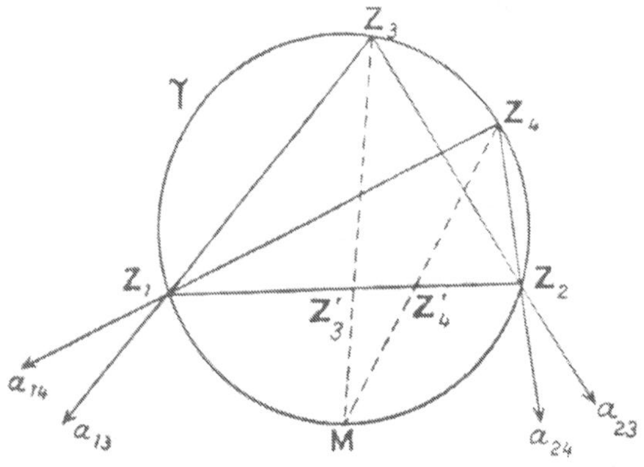

Fig. 17

Points Z_3, Z_4 may or may not belong to a common arc defined by the points Z_1, Z_2.

If M is the midpoint of the arc which does not contain Z_3, the lines MZ_3, MZ_4 are bisectors of the angles $(a_{23}a_{13})$, (a_{24},a_{14}) and intersect the line Z_1Z_2 in Z'_3, Z'_4. We have, in both magnitude and sign, for each of the two figures,

$$\frac{Z_1Z_3}{Z_2Z_3} = -\frac{Z_1Z'_3}{Z_2Z'_3},$$

$$\frac{Z_1Z_4}{Z_2Z_4} = -\frac{Z_1Z'_4}{Z_2Z'_4},$$

and equation (1) becomes

$$(Z_1Z_2Z_3Z_4) = (Z_1Z_2Z'_3Z'_4).$$

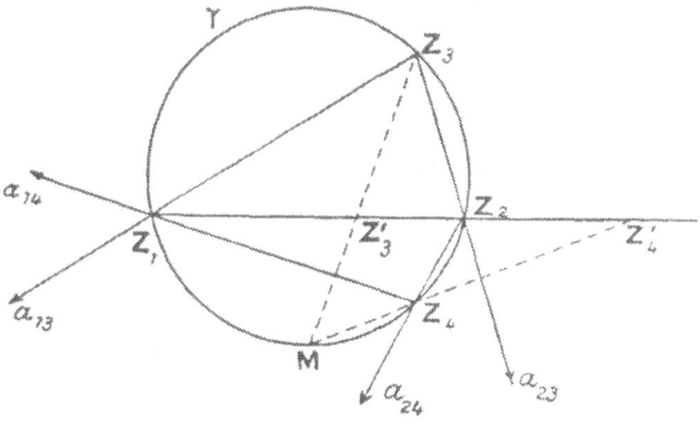

FIG. 18

But, by projecting the second A.R. from point M, we conclude that it is equal to the A.R. defined in modern plane geometry for the four points Z_1, Z_2, Z_3, Z_4 of the circle γ.

Conversely, if the points Z_1, Z_2, Z_3, Z_4 are collinear or concyclic, we have equation (2), and equation (1) proves that the A.R. $(Z_1Z_2Z_3Z_4)$ is real.

29. Construction. *Being given three points* Z_1, Z_2, Z_3, *as well as a real or imaginary number of modulus* r *and argument* θ, *there exists a unique point* Z_4 *such that*

$$(Z_1Z_2Z_3Z_4) = re^{i\theta} \tag{3}$$

and we shall construct it.

From the equation

$$(z_1z_2z_3z_4) = \frac{z_1 - z_3}{z_2 - z_3} : \frac{z_1 - z_4}{z_2 - z_4} = re^{i\theta}$$

we obtain

$$z_4\,[(z_2\text{-}z_3)re^{i\theta} - (z_1\text{-}z_3)] = z_1(z_2\text{-}z_3)re^{i\theta} - z_2(z_1\text{-}z_3). \tag{4}$$

In this equation, which is linear in z_4, we cannot have at the same time

$$(z_2\text{-}z_3)re^{i\theta} - (z_1\text{-}z_3) = 0, \tag{5}$$
$$z_1(z_2\text{-}z_3)re^{i\theta} - z_2(z_1\text{-}z_3) = 0,$$

for this would make

$$\begin{vmatrix} z_2\text{-}z_3 & z_1\text{-}z_3 \\ z_1(z_2\text{-}z_3) & z_2(z_1\text{-}z_3) \end{vmatrix} = (z_2\text{-}z_3)(z_1\text{-}z_3)(z_2\text{-}z_1) = 0,$$

which is impossible since the points Z_1, Z_2, Z_3 are supposed to be distinct.

If (5) does not hold, then equation (4) has the unique finite solution

$$z_4 = \frac{z_1(z_2 - z_3)re^{i\theta} - z_2(z_1 - z_3)}{(z_2 - z_3)re^{i\theta} - (z_1 - z_3)}, \tag{6}$$

which is the affix of the only point Z_4 fulfilling the requirement.

If (5) does hold, that is, if

$$\frac{z_1 - z_3}{z_2 - z_3} = re^{i\theta}$$

or (9)

$$\left|\frac{Z_3Z_1}{Z_3Z_2}\right| = r, \quad (\overline{Z_3Z_2}, \overline{Z_3Z_1}) = \theta,$$

z_4 is infinite and the point at infinity of the Gauss plane is the only point Z_4 which fulfills the requirement.

To construct Z_4 when (5) does not hold, we do not employ the expression (6) for z_4, but we write equation (3) in the form (25)

$$\left|\frac{Z_1Z_3}{Z_2Z_3} : \frac{Z_1Z_4}{Z_2Z_4}\right| e^{i[(\overline{Z_3Z_2}, \overline{Z_3Z_1}) - (\overline{Z_4Z_2}, \overline{Z_4Z_1})]} = re^{i\theta}.$$

By equating the moduli and then the arguments of these two complex numbers we obtain

$$\left|\frac{Z_1Z_4}{Z_2Z_4}\right| = \frac{1}{r}\left|\frac{Z_1Z_3}{Z_2Z_3}\right|, \tag{7}$$

$$(\overline{Z_4Z_2}, \overline{Z_4Z_1}) = (\overline{Z_3Z_2}, \overline{Z_3Z_1}) - \theta. \tag{8}$$

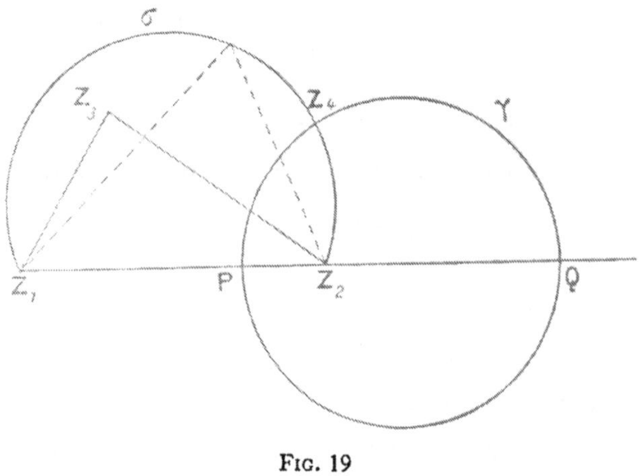

FIG. 19

Equation (7) says that point Z_4 is on the circle γ which is the locus of points the ratio of whose distances from points Z_1, Z_2 is

$$r_1 = \frac{1}{r}\left|\frac{Z_1Z_3}{Z_2Z_3}\right|;$$

this circle has for diameter the segment PQ of the line Z_1Z_2, where P and Q are defined by

$$\frac{Z_1P}{Z_2P} = -r_1, \quad \frac{Z_1Q}{Z_2Q} = r_1$$

and are thus harmonic conjugates with respect to Z_1, Z_2. In the case where $r_1 = 1$, that is to say, if $r = |Z_1Z_3/Z_2Z_3|$, circle γ is replaced by the perpendicular bisector of the segment Z_1Z_2.

Equation (8) says that Z_4 is on the arc σ described on Z_2Z_1 and for which Z_2Z_1 subtends an angle of algebraic value

$$\theta = (\overline{Z_3Z_2}, \overline{Z_3Z_1}) - \theta,$$

whose sign specifies on which side of the line Z_1Z_2 in the oriented plane we should draw σ. In the case where θ_1 is an integral multiple of π, the line Z_1Z_2 replaces σ.

Point Z_4 is the intersection of γ and σ. It will be noticed that Z_4 is at infinity if $r_1 = 1$, $\theta_1 = 0$; it is the midpoint of the segment Z_1Z_2 if $r_1 = 1$, $\theta_1 = \pi$, and in the two cases, γ and σ are lines.

30. Harmonic quadrangle. When the A.R. of four points Z_1, Z_2, Z_3, Z_4 of the complex plane is equal to -1, it is said to be harmonic.

The four points are necessarily on a line or on a circle (28). In either case, if

$$(Z_1Z_2Z_3Z_4) = -1, \tag{9}$$

we say that Z_1, Z_2 are harmonic conjugates with respect to Z_3, Z_4 or harmonically separate Z_3, Z_4, and also, since $(Z_3Z_4Z_1Z_2) = -1$ (26), that Z_3, Z_4 harmonically separate Z_1, Z_2. The figure formed by the four points is called a *harmonic quadrangle*, and is degenerate if the points are collinear.

I. *The harmonic conjugate Z_4 of Z_3 with respect to Z_1, Z_2 is*

1° *the second point common to the circle O circumscribed about triangle $Z_1Z_2Z_3$ and the circle of Apollonius O_3 associated with vertex Z_3;*

2° *the second point common to circle O and the symmedian of triangle $Z_1Z_2Z_3$ drawn through vertex Z_3; this line joins Z_3 to the pole P_3 of side Z_1Z_2 in circle O and is symmetric to the median Z_3M_3 with respect to the bisector Z_3N_3 of angle $Z_1Z_3Z_2$.*

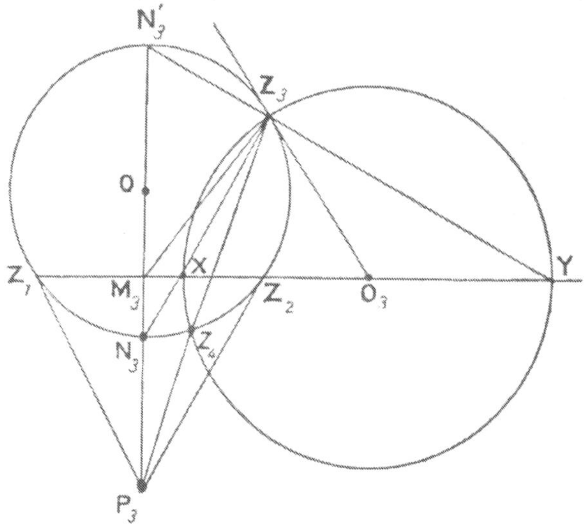

FIG. 20

Taking the moduli of the two members of equation (9) we obtain

(25)
$$\left|\frac{Z_1Z_4}{Z_2Z_4}\right| = \left|\frac{Z_1Z_3}{Z_2Z_3}\right|$$

and Z_4 is on the circle which is the locus of points the ratio of whose distances from Z_1, Z_2 is that of the distances of Z_3 from Z_1, Z_2. This circle passes through Z_3 and through the intersections X, Y with line Z_1Z_2 of the bisectors Z_3N_3, Z_3N_3' of angle $Z_1Z_3Z_2$, where N_3, N_3' denote the intersections of circle O with the diameter normal to Z_1Z_2; this is the circle of Apollonius of triangle $Z_1Z_2Z_3$ associated with vertex Z_3, and its center O_3 is the midpoint of segment XY inasmuch as angle XZ_3Y is a right angle.

Since $(Z_1Z_2XY) = -1$, the circle of Apollonius O_3 is orthogonal to circle O and the line Z_3Z_4 is the polar of point O_3 with respect to circle O and thus contains the pole P_3 of the line Z_1Z_2.

From $(P_3M_3N_3N_3') = -1$ we obtain the harmonic pencil $Z_3(P_3M_3N_3N_3')$ in which the conjugate rays Z_3N_3, Z_3N_3' are perpendicular and are therefore the bisectors of the angles formed by the other two rays Z_3P_3, Z_3M_3, whence Z_3P_3 is the symmedian drawn through vertex Z_3 in triangle $Z_1Z_2Z_3$.

It will be noticed that from the graphic point of view the point Z_4 is determined more neatly by the circle O_3 than by the line Z_3P_3, for circle O is cut orthogonally by circle O_3 but not by line Z_3P_3 except when Z_3 is at N_3 or N_3'.

II. *If* $(Z_1Z_2Z_3Z_4) = -1$, *we have the equations*

$$(z_1 + z_2)(z_3 + z_4) = 2(z_1z_2 + z_3z_4), \tag{10}$$

$$\frac{2}{z_1 - z_2} = \frac{1}{z_1 - z_3} + \frac{1}{z_1 - z_4} \tag{11}$$

and conversely.

Equation $(z_1z_2z_3z_4) = -1$ is equivalent to (26, 5°)

$$\frac{\dfrac{1}{z_1 - z_2} - \dfrac{1}{z_1 - z_4}}{\dfrac{1}{z_1 - z_2} - \dfrac{1}{z_1 - z_3}} = -1$$

which is easily written in the form (11), and then, by clearing of fractions, as equation (10).

Corollary. *The affix of the midpoint* Z_3 *of a segment* Z_1Z_2 *is half the sum of the affixes of* Z_1, Z_2.

In fact we have (27)

$$-1 = (Z_1Z_2Z_3\,\infty) = (z_1z_2z_3\,\infty) = \frac{z_1 - z_3}{z_2 - z_3}, \quad z_3 = \frac{1}{2}(z_1 + z_2).$$

III. *If* $(Z_1Z_2Z_3Z_4) = -1$ *and if* M_3 *is the midpoint of the segment* Z_1Z_2 *joining two conjugates, we have*

$$(m_3 - z_1)^2 = (m_3 - z_3).(m_3 - z_4), \tag{12}$$

$$M_3Z_1^2 = |\, M_3Z_3.M_3Z_4 \,| \tag{13}$$

and the line Z_1Z_2 *is the interior bisector of angle* $(\overline{M_3Z_3}, \overline{M_3Z_4})$.

From $(z_1z_2z_3z_4) = -1$ we obtain the proportion

$$\frac{z_1 - z_3}{z_2 - z_3} = \frac{z_4 - z_1}{z_2 - z_4}$$

from which we get, since the sum of the two first terms is to their difference as the sum of the two last terms is to their difference,

$$\frac{z_1 + z_2 - 2z_3}{z_1 - z_2} = \frac{z_2 - z_1}{2z_4 - (z_1 + z_2)}. \tag{14}$$

If we observe that (II, *corollary*)

$$z_1 + z_2 = 2\,m_3, \qquad z_2 - z_1 = 2\,(m_3 - z_1),$$

equation (14) takes the form (12).

The equality of the moduli of the two members of (12) yields equation (13).

If we place the origin of cartesian coordinates at M_3 and the x-axis on Z_1Z_2, equation (12) becomes

$$z_1^2 = z_3 z_4.$$

Since z_1^2 is a positive real number, its argument is zero, which is then also the sum of the arguments of z_3, z_4, and the proof is complete.

Conversely, if we have equation (12), *then point* M_3 *is the midpoint of the segment joining* Z_1 *to its harmonic conjugate* Z_2 *with respect to points* Z_3, Z_4.

In fact, (12) can be written as

$$m_3\,(2z_1 - z_3 - z_4) = z_1^2 - z_3 z_4$$

which is linear in m_3. The value of m_3 is that of the affix z of the midpoint of Z_1Z_2 if $(Z_1Z_2Z_3Z_4) = -1$ since we must have

$$(z - z_1)^2 = (z - z_3)\,(z - z_4) \quad \text{or} \quad z\,(2z_1 - z_3 - z_4) = z_1^2 - z_3 z_4.$$

31. Construction problems. 1. *Being given three distinct points* Z_3, Z_4, M_3, *to construct in the Gauss plane two points* Z_1, Z_2 *which harmonically separate* Z_3, Z_4 *and such that* M_3 *shall be the midpoint of the segment* Z_1Z_2.

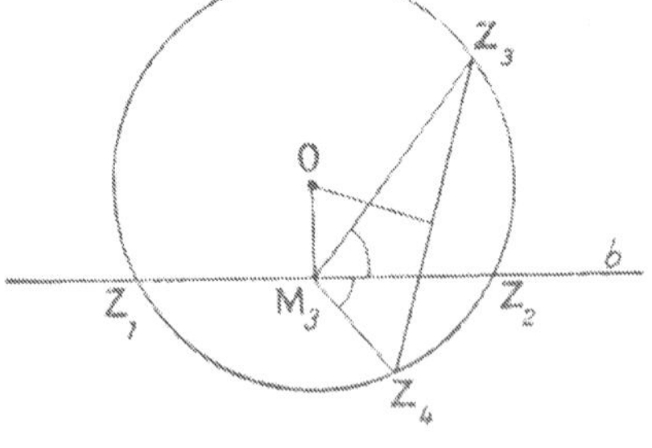

FIG. 21

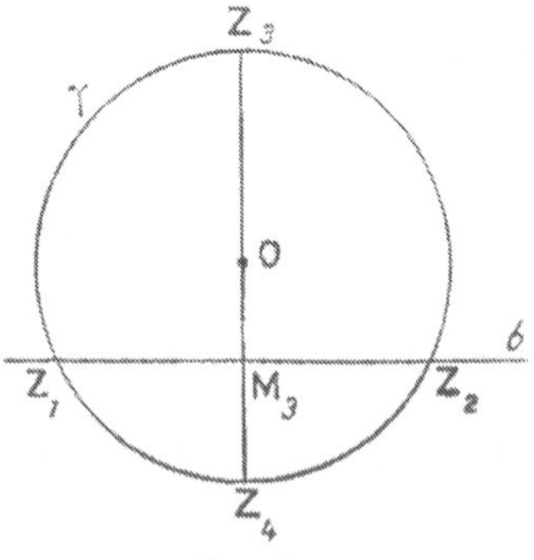

FIG. 22

If Z_3, Z_4, M_3 are not collinear (Fig. 21), the sought points are (30, III) the intersections of the interior bisector b of angle $(\overline{M_3Z_3}, \overline{M_3Z_4})$ with the circle γ which passes through Z_3, Z_4 and whose center O is on the perpendicular to b at M_3.

The construction is still applicable if Z_3, Z_4, M_3 are collinear, with M_3 between Z_3 and Z_4 (Fig. 22). It does not succeed if M_3 is on an extension of the segment Z_3Z_4, for then b coincides with the line Z_3Z_4; in this case b contains Z_1, Z_2, and these points are determined by the equation $M_3Z_1^2 = M_3Z_3.M_3Z_4$ (Fig. 23).

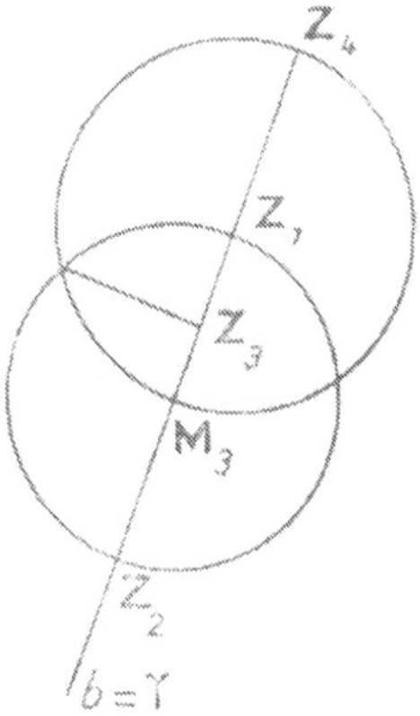

FIG. 23

2. *Being given the real or imaginary numbers* a, b, c, *to construct the points* Z *whose affixes* z *are the roots of the linear equations*

$$\frac{2}{z-a} = \frac{1}{z-b} + \frac{1}{z-c}, \tag{15}$$

$$\frac{2}{z} = \frac{1}{a} + \frac{1}{b}, \tag{16}$$

$$\frac{1}{z} = \frac{1}{a} + \frac{1}{b}, \tag{17}$$

$$\frac{1}{z} = \frac{1}{a_1} + \frac{1}{a_2} + \dots + \frac{1}{a_n}. \tag{18}$$

The form of equation (15) shows that Z is the harmonic conjugate of the point A, of affix a, with respect to the points B, C, of affixes b, c (**30**, **II**), and can be constructed by the procedure indicated above (**30**, **I**).

Equation (16) written in the form

$$\frac{2}{o-z} = \frac{1}{o-a} + \frac{1}{o-b}$$

shows that Z is the harmonic conjugate of the origin O with respect to the points A, B of affixes a, b.

If in equation (17) we set

$$z_1 = 2z,$$

it becomes

$$\frac{2}{z_1} = \frac{1}{a} + \frac{1}{b}.$$

After constructing point Z_1, the harmonic conjugate of O with respect to the points A, B of affixes a, b, we obtain Z as the midpoint of OZ_1 (8).

As for equation (18), we can construct successively the points Z_2, Z_3, ..., Z from

$$\frac{1}{z_2} = \frac{1}{a_1} + \frac{1}{a_2}, \quad \frac{1}{z_3} = \frac{1}{z_2} + \frac{1}{a_3}, \quad \dots, \quad \frac{1}{z} = \frac{1}{z_{n-1}} + \frac{1}{a_n}.$$

A second procedure consists in constructing the points A_1, A_2, ..., A_n of affixes $a_1, a_2, \dots, a_n$, then their inverses A'_1, A'_2, ..., A'_n under the inversion of center O and power 1, and having for affixes $1/\bar{a}_1$, $1/\bar{a}_2$, ..., $1/\bar{a}_n$ (9) ; if we construct the point $\bar{Z}'$ of affix

$$\bar{z}' = \frac{1}{\bar{a}_1} + \frac{1}{\bar{a}_2} + \dots + \frac{1}{\bar{a}_n},$$

we have

$$z' = \frac{1}{z}, \qquad z = \frac{1}{z'}$$

and the sought point Z is consequently the inverse of point $\bar{Z}'$.

3. *Being given three distinct numbers* a, b, c, *as well as three non-zero numbers* α, β, γ *whose sum is zero, to construct the point* Z *defined by*

$$\frac{\alpha}{z-a} + \frac{\beta}{z-b} + \frac{\gamma}{z-c} = 0. \qquad (19)$$

We are going to show that *if* A, B, C *are the points of affixes* a, b, c, *we have*

$$(ABCZ) = -\frac{\beta}{\alpha},$$

which permits us to construct Z as has been shown in article **29.**

From the equation

$$\alpha + \beta + \gamma = 0$$

we find, if we set

$$-\frac{\beta}{\alpha} = \lambda,$$

$$\frac{\gamma}{\alpha} = -1 + \lambda$$

and equation (19) yields successively

$$\frac{1}{z-a} - \frac{\lambda}{z-b} + \frac{\lambda - 1}{z-c} = 0,$$

$$\lambda = \frac{\dfrac{1}{z-c} - \dfrac{1}{z-a}}{\dfrac{1}{z-c} - \dfrac{1}{z-b}}.$$

We therefore have (**26**)

$$\lambda = (\mathrm{ZCBA}) = (\mathrm{ABCZ}).$$

4. *If* a, b, c *are three distinct numbers, to construct the point* Z *such that*

$$(z-a)^2 = (z-b)(z-c).$$

If A, B, C are the points of affixes a, b, c, point Z is the midpoint of the segment which joins A to its harmonic conjugate with respect to B, C (**30, III**).

32. Equianharmonic quadrangle. When four points constitute a harmonic quadrangle (**30**), the principal A.R.'s are -1, $1/2$, 2, and the 24 possible A.R.'s separate themselves into three sets of eight ratios having these values, for these have -1, 2, $1/2$ for their reciprocals.

We are going to show that there exist in the Gauss plane quadruples of points for which the 24 A.R.'s lead to only two values, instead of the 6 of the general case and the 3 in the case of a harmonic quadruple.

If four points are such that two principal A.R.*'s are equal, they are equal to a third and have for value either* $e^{i\pi/3}$ *or* $e^{-i\pi/3}$. *We say that the anharmonic ratio is* **equianharmonic** *and that the four points form an* **equianharmonic quadrangle**; 12 *of their* A.R.*'s have the value* $e^{i\pi/3}$, *the other* 12 *have the value* $e^{-i\pi/3}$.

The principal A.R.'s of any four points have the values (**26**)

$$\lambda, \quad \frac{1}{1-\lambda}, \quad \frac{\lambda-1}{\lambda}.$$

If we have

$$\frac{\lambda}{1} = \frac{1}{1-\lambda}, \tag{20}$$

we obtain, by taking the difference of the antecedents and of the consequents,

$$\lambda = \frac{1}{1-\lambda} = \frac{\lambda-1}{\lambda}.$$

We arrive at the same result if we start from either of the relations

$$\lambda = \frac{\lambda-1}{\lambda}, \quad \frac{1}{1-\lambda} = \frac{\lambda-1}{\lambda}.$$

Equation (20) gives

$$\lambda^2 - \lambda + 1 = 0,$$

$$\lambda = \frac{1}{2} \pm i\frac{\sqrt{3}}{2} = \cos\frac{\pi}{3} \pm i\sin\frac{\pi}{3} = e^{\pm i\pi/3}.$$

Properties. 1° *Given any three points* Z_1, Z_2, Z_3 *in the Gauss plane, there exist two points* W, W′ *which form with the given points two equianharmonic quadrangles.*

The points W, W′, *called the* **isodynamic centers** *of triangle* $Z_1Z_2Z_3$, *are common to the three circles of Apollonius of the triangle.*

In an equianharmonic quadrangle, the three pairs of opposite sides have equal products, and conversely.

We know (29) that there exists a unique point W such that

$$(Z_1Z_2Z_3W) = e^{i\pi/3}$$

and a unique point W′ such that

$$(Z_1Z_2Z_3W') = e^{-i\pi/3}$$

and we are able to construct these points by noticing, for W for example, that

$$\left|\frac{Z_1Z_3}{Z_2Z_3} : \frac{Z_1W}{Z_2W}\right| = 1, \quad (\overline{Z_3Z_2}, \overline{Z_3Z_1}) - (\overline{WZ_2}, \overline{WZ_1}) = \frac{\pi}{3}.$$

But we can determine the pair W, W′ by considering only the moduli, if we write

$$(Z_1Z_2Z_3W) = (Z_1Z_3WZ_2) = (Z_1WZ_2Z_3) = e^{i\pi/3},$$

for from this we obtain

$$\left|\frac{Z_1W}{Z_2W}\right| = \left|\frac{Z_1Z_3}{Z_2Z_3}\right|, \quad \left|\frac{Z_1W}{Z_3W}\right| = \left|\frac{Z_1Z_2}{Z_3Z_2}\right|, \quad \left|\frac{Z_3W}{Z_2W}\right| = \left|\frac{Z_3Z_1}{Z_2Z_1}\right|$$

and similar equalities for W′, whence W, W′ are common to the three circles of Apollonius.

The preceding equations give

$$|Z_1W.Z_2Z_3| = |Z_2W.Z_3Z_1| = |Z_3W.Z_1Z_2|$$

and conversely.

2° *In order that a triangle* $Z_1Z_2Z_3$ *be equilateral, it is necessary and sufficient that one of the isodynamic centers be the point at infinity in the Gauss plane; the other isodynamic center is the point of concurrency of the medians of the triangle.*

The condition is necessary. In fact, if $Z_1Z_2Z_3$ is equilateral, we have

$$|Z_1Z_3| = |Z_2Z_3|, \qquad (\overline{Z_3Z_2}, \overline{Z_3Z_1}) = \pm \frac{\pi}{3} \tag{21}$$

and consequently (**25, 27**)

$$(Z_1Z_2Z_3\infty) = e^{\pm i\pi/3}, \tag{22}$$

from which it follows that the point at infinity is an isodynamic center. Since the circles of Apollonius reduce to the perpendicular bisectors of the sides, the second isodynamic center is the point mentioned.

The condition is sufficient, for equation (22) carries with it equations (21), and these express that the triangle is equilateral.

3° *If the points* Z_1, Z_2, Z_3 *are collinear, and if we construct* Z'_1, Z'_2, Z'_3 *such that*

$$(Z_2Z_3Z_1Z'_1) = -1, \quad (Z_3Z_1Z_2Z'_2) = -1, \quad (Z_1Z_2Z_3Z'_3) = -1,$$

then the isodynamic centers of the triple Z_1, Z_2, Z_3 *are the points* W, W′ *common to the three circles having for diameters the segments* $Z_1Z'_1$, $Z_2Z'_2$, $Z_3Z'_3$.

These circles are, in fact, for the degenerate triangle $Z_1Z_2Z_3$, the circles of Apollonius associated with the vertices Z_1, Z_2, Z_3.

Exercises 17 through 31

17. Images of the square roots of a number. The images A_1, A_2 of the square roots a_1, a_2 of a number a with image A are harmonic conjugates with respect to A and the point U of affix 1, and the segment A_1A_2 has the origin O for midpoint. We can construct A_1, A_2 by the process of article **31, 1.** [a_1, a_2 are roots of $z^2 = a$, which can be written as

$$(0 - z)^2 = (0 - a)(0 - 1).$$

See article **31, 4.**]

Construct the images of the square roots, then of the fourth roots, of i, $-i$, $1 + i$.

18. Images of the roots of the quadratic equation

$$z^2 - pz + q = 0.$$

Construct the point M of affix $p/2$ and the images Q_1, Q_2 of the square roots of q. The images Z_1, Z_2 of the roots of the equation harmonically separate Q_1, Q_2 and M is the midpoint of Z_1Z_2. See article **31** and exercise 17. [Write the equation as

$$\left(\frac{p}{2} - z\right)^2 = \left(\frac{p}{2} - q_1\right)\left(\frac{p}{2} - q_2\right).$$

Another procedure, generally less simple, is to translate the classical formula

$$\frac{p}{2} \pm \sqrt{\frac{p^2}{4} - q}$$

by first constructing the image of $p^2/4 - q$.]

Deduce a construction of the images of the roots of the quartic equation

$$z^4 - pz^2 + q = 0.$$

19. Two points Z_1, Z_2 harmonically separate the images of the roots of

$$az^2 + 2bz + c = 0$$

if we have

$$az_1z_2 + b(z_1 + z_2) + c = 0.$$

[Apply II of article 30.]

20. The images Z_1, Z_2 of the roots of the equation

$$az^2 + 2bz + c = 0$$

harmonically separate the images Z_1', Z_2' of the roots of

$$a'z^2 + 2b'z + c' = 0$$

if we have

$$ac' - 2bb' + ca' = 0.$$

21. Pair E, F harmonically separating two given pairs Z_1, Z_2 and Z_3, Z_4. By exercise 19, the affixes of E, F are the roots of the equation in z

$$\begin{vmatrix} z^2 & 2z & 1 \\ z_1z_2 & z_1 + z_2 & 1 \\ z_3z_4 & z_3 + z_4 & 1 \end{vmatrix} = 0.$$

The construction of E, F reduces (article 31) to that of the midpoint M of segment EF. The equation gives

$$m = \frac{z_1z_2 - z_3z_4}{z_1 + z_2 - z_3 - z_4}.$$

Since

$$2(z_1z_2 - z_3z_4) = (z_1 + z_3)(z_2 + z_3) + (z_1 + z_4)(z_2 + z_4) - (z_3 + z_4)(z_1 + z_2 + z_3 + z_4),$$

if M_{ik} is the midpoint of Z_iZ_k and G the common midpoint of the segments $M_{12}M_{34}$, $M_{13}M_{24}$, $M_{14}M_{23}$, we have

$$m = \frac{m_{13}m_{23} + m_{14}m_{24} - 2m_{34}g}{m_{12} - m_{34}}.$$

If we take the origin of axes at G, we have

$$g = 0, \qquad m_{24} = -m_{13}, \qquad m_{14} = -m_{23}, \qquad m_{34} = -m_{12}, \qquad m = \frac{m_{13}m_{23}}{m_{12}}.$$

Therefore, if M_{12}, M_{23}, M_{13} are the midpoints of the segments Z_1Z_2, Z_2Z_3, Z_3Z_1 and if G is the barycenter of the points Z_1, Z_2, Z_3, Z_4, vector $\overline{GM}$ has a sense symmetric to that of vector $\overline{GM_{12}}$ with respect to the interior bisector of angle $(\overline{GM_{13}},\overline{GM_{23}})$ and

$$|GM| = \frac{|GM_{13}|\,|GM_{23}|}{|GM_{12}|}.$$

For another construction of M, see article **104**, where M is denoted by O.

22. If the affixes of two pairs of points are the roots of

$$a_1z^2 + 2b_1z + c_1 = 0 \qquad \text{and} \qquad a_2z^2 + 2b_2z + c_2 = 0,$$

then the points which harmonically separate each of these pairs are given by

$$(a_2b_1 - a_1b_2)z^2 + (a_2c_1 - a_1c_2)z + b_2c_1 - b_1c_2 = 0.$$

23. If three pairs of points are such that each pair harmonically separates the other two, then the midpoints of any two pairs harmonically separate the third pair.

24. Show that, for any point M, we have

$$(ABCD) = (ABCM)\,(ABMD).$$

From this show that, in order to change only the sign of an anharmonic ratio (ABCD), it suffices to replace one point in one of the pairs (A,B), (C,D) by its harmonic conjugate with respect to the other pair. Thus, if

$$(AA'CD) = -1,$$

we have

$$(A'BCD) = -(ABCD).$$

25. Being given three points A, B, C, we construct the harmonic conjugate of each of them with respect to the other two, so that

$$(AA'BC) = -1, \qquad (BB'CA) = -1, \qquad (CC'AB) = -1.$$

By using the properties of article **26** and of exercise 24, show that:

1° $(ABCA') = 1/2$, $(ABCB') = 2$;

2° the elimination of C gives

$$(ABA'B') = 4 \qquad \text{and} \qquad (AA'BB') = (BB'CC') = (CC'AA') = -3;$$

3° from 1° we obtain

$$(ABC'A') = -1/2, \qquad (ABC'B') = -2,$$

and, by eliminating A,

$$(BB'C'A') = (CC'A'B') = (AA'B'C') = -1,$$

whence A is the harmonic conjugate of A' with respect to B', C', *etc.*

26. In exercise 25, designate the affixes of A, B, C by a, b, c and set

$$\sigma_1 = a + b + c, \qquad \sigma_2 = ab + bc + ca, \qquad \sigma_3 = abc.$$

By first of all calculating the affixes a', b', c' of A′, B′, C′ as functions of a, b, c (article **30**, II), show that:

1° $$\frac{1}{a - a'} + \frac{1}{b - b'} + \frac{1}{c - c'} = 0;$$

2° $$a' = \frac{3\sigma_3 - a\sigma_2}{a\sigma_1 - 3a^2} \tag{1}$$

and that, a, b, c being the roots of the cubic equation

$$z^3 - \sigma_1 z^2 + \sigma_2 z - \sigma_3 = 0, \tag{2}$$

a', b', c' are roots of

$$pz^3 - 3qz^2 - 3rz + s = 0,$$

where

$$p = 2\sigma_1^3 - 9\sigma_1\sigma_2 + 27\sigma_3, \qquad q = \sigma_1^2\sigma_2 + 9\sigma_1\sigma_3 - 6\sigma_2^2,$$

$$r = \sigma_1\sigma_2^2 + 9\sigma_2\sigma_3 - 6\sigma_1^2\sigma_3, \qquad s = 2\sigma_2^3 - 9\sigma_1\sigma_2\sigma_3 + 27\sigma_3^2;$$

[Set $a = z$ in (1) and then eliminate z from (1) and (2).]

3° there exist two points E, F which harmonically separate each of the pairs AA′, BB′, CC′ (see exercise 19) and their affixes are the roots of

$$(3\sigma_2 - \sigma_1^2)z^2 + (\sigma_1\sigma_2 - 9\sigma_3)z + 3\sigma_1\sigma_3 - \sigma_2^2 = 0. \tag{3}$$

27. The isodynamic centers W, W′ of the system of points A, B, C are the points E, F of 3° of exercise 26. [Show (article **32**) that

$$(abcw)^2 - (abcw) + 1 = 0$$

and find that w is a root of equation (3) of exercise 26.]

28. The images of the roots of the equation

$$a_0 z^4 + 4a_1 z^3 + 6a_2 z^2 + 4a_3 z + a_4 = 0$$

form two harmonically separating pairs if

$$\begin{vmatrix} a_0 & a_1 & a_2 \\ a_1 & a_2 & a_3 \\ a_2 & a_3 & a_4 \end{vmatrix} = 0,$$

and the affixes of the midpoints of the pairs are

$$-\frac{a_1}{a_0} \pm \frac{1}{a_0}\sqrt{a_1^2 - a_0 a_2}.$$

If $a_1^2 = a_0 a_2$, the two pairs are the vertices of a square whose center has affix $-a_1/a_0$.

[If

$$z_1 + z_2 = t, \qquad z_3 + z_4 = u, \qquad z_1 z_2 = v, \qquad z_3 z_4 = w,$$

the harmonic relation and the elementary functions of the roots yield tu, $t + u$, vw, $v + w$, $vu + wt$, $wu + vt$, and then

$$(vu + wt)(wu + vt)$$

leads to the desired result.]

29. A necessary and sufficient condition for the images of the roots of the sixth degree equation

$$a_0 z^6 + 6a_1 z^5 + 15a_2 z^4 + 20a_3 z^3 + 15a_4 z^2 + 6a_5 z + a_6 = 0$$

to be separable into three pairs such that each pair harmonically separates the other two is that we have

$$a_4 = \frac{4a_1 a_3 - 3a_2^2}{a_0},$$

$$a_5 = \frac{12a_1^2 a_3 - 9a_1 a_2^2 - 2a_0 a_2 a_3}{a_0^2},$$

$$a_6 = \frac{36a_1 a_2 a_3 - 27a_2^3 - 8a_0 a_3^2}{a_0^2}.$$

The affixes of the midpoints of the pairs are roots of

$$a_0 z^3 + 3a_1 z^2 + 3a_2 z + a_3 = 0$$

and the products of the affixes of the points of each pair are roots of

$$a_0 z^3 - 3a_2 z^2 + 3a_4 z - a_6 = 0.$$

[Set

$$z_1 + z_2 = 2u, \qquad z_3 + z_4 = 2v, \qquad z_5 + z_6 = 2w,$$

$$z_1 z_2 = \mathrm{U}, \qquad z_3 z_4 = \mathrm{V}, \qquad z_5 z_6 = \mathrm{W}.$$

The symmetric functions of the roots are

$$\Sigma u, \quad \Sigma \mathrm{U} + 4\Sigma uv, \quad \Sigma(v + w)\mathrm{U} + 4uvw, \quad \Sigma \mathrm{UV} + 4\Sigma vw\mathrm{U}, \quad \Sigma u\mathrm{VW}, \quad \mathrm{UVW},$$

to which we add

$$2uv = \mathrm{U} + \mathrm{V},$$

..., which express the harmonic relations. From these obtain U, V, W, to put in the preceding functions, whence

$$\Sigma uv = \frac{3a_2}{a_0}, \qquad uvw = -\frac{a_3}{a_0}, \ \ldots \ .]$$

30. A necessary and sufficient condition for the existence of two points E, F harmonically separating each of the three pairs given by

$$a_0z^2 + 2a_1z + a_2 = 0, \qquad b_0z^2 + 2b_1z + b_2 = 0, \qquad c_0z^2 + 2c_1z + c_2 = 0,$$

is that

$$\begin{vmatrix} a_0 & a_1 & a_2 \\ b_0 & b_1 & b_2 \\ c_0 & c_1 & c_2 \end{vmatrix} = 0.$$

All pairs harmonically separating E, F are given by

$$\lambda(a_0z^2 + 2a_1z + a_2) + \mu(b_0z^2 + 2b_1z + b_2) = 0,$$

where λ and μ are arbitrary.

31. Does there exist a point P such that, given two triples of points A, B, C and A′, B′, C′, the anharmonic ratios (ABCP), (A′B′C′P) have unit moduli? Construct P.

CHAPTER TWO

ELEMENTS OF ANALYTIC GEOMETRY IN COMPLEX NUMBERS

I. GENERALITIES

33. Passage to complex coordinates. Let x and y be the cartesian coordinates and z the complex coordinate, relative to two perpendicular axes Ox and Oy, of a real point Z. We have (1)

$$x + iy = z$$

and consequently

$$x - iy = \bar{z}.$$

From these two equations we obtain the formulas

$$\boxed{x = \frac{1}{2}(z + \bar{z}), \qquad y = \frac{1}{2i}(z - \bar{z})}$$

permitting us *to pass from the rectangular cartesian coordinates to the complex coordinate.*

The polar coordinates θ and r of Z, relative to the pole O and the polar axis Ox, are related to x and y by the equations (3)

$$x = r\cos\theta, \qquad y = r\sin\theta$$

from which we obtain

$$r^2 = x^2 + y^2 = (x + iy)(x - iy) = z\bar{z},$$

$$\tan\theta = \frac{y}{x} = \frac{z - \bar{z}}{i(z + \bar{z})}.$$

We therefore pass from the polar coordinates to the complex coordinate by the formulas

$$r = (z\bar{z})^{\frac{1}{2}}, \qquad \theta = \text{arc tan}\,\frac{z - \bar{z}}{i(z + \bar{z})}.$$

34. Parametric equation of a curve. If, relative to two perpendicular axes Ox, Oy, a curve c has parametric equations

$$x = f_1(t), \qquad y = f_2(t), \tag{1}$$

$f_1(t)$, $f_2(t)$ being real functions of a real parameter t, the complex parametric equation of c is

$$z = x + iy = f_1(t) + i f_2(t)$$

or

$$z = f(t), \tag{2}$$

$f(t)$ being a complex function of the real parameter t.

Conversely, we pass from (2) to (1) by separating the imaginary and the real parts.

Direction components of the tangent to c at the point with parameter t being

$$\frac{dx}{dt} = f_1'(t), \qquad \frac{dy}{dt} = f_2'(t),$$

the complex number

$$\frac{dx}{dt} + i\frac{dy}{dt} \quad \text{or} \quad \frac{dz}{dt}$$

is the affix of a point which, joined to O, gives a line parallel to the considered tangent. Therefore:

If $z = f(t)$ *is the complex parametric equation of a curve* c, *the tangent at the point with parameter* t *is parallel to the line joining the origin to the point of affix* dz/dt.

II. THE STRAIGHT LINE

35. Point range formula. *If* z_1, z_2, z *are the affixes of two points* Z_1, Z_2 *and of the point* Z *which divides the segment* Z_1Z_2 *in the ratio*

$$k = \frac{Z_1Z}{ZZ_2},$$

we have

$$z = \frac{z_1 + kz_2}{1 + k}. \tag{1}$$

By analytic geometry, we know that if (x_1, y_1), (x_2, y_2), (x, y) are the rectangular cartesian coordinates of Z_1, Z_2, Z, then

$$x = \frac{x_1 + kx_2}{1 + k}, \quad y = \frac{y_1 + ky_2}{1 + k}.$$

Since

$$z = x + iy, \qquad z_1 = x_1 + iy_1, \qquad z_2 = x_2 + iy_2,$$

we obtain equation (1). This result also follows from the classic vector equation

$$\overline{OZ} = \frac{\overline{OZ_1} + k\overline{OZ_2}}{1 + k}.$$

Corollary. By taking $k = 1$, we again find (30, II) the *affix*

$$z = \tfrac{1}{2}(z_1 + z_2)$$

of the midpoint of the segment Z_1Z_2.

36. Parametric equation. *If a line passes through the point* A *of affix* a *and is parallel to the line joining the origin to the point* B *of affix* b, *its parametric equation is*

$$\boxed{z = a + bt,} \qquad (2)$$

where t *denotes a real parameter which can vary from* $-\infty$ *to* $+\infty$.

If Z is any point whatever of the line, we have

$$\overline{OZ} = \overline{OA} + \overline{AZ}$$

or, since

$$\overline{AZ} = t\,\overline{OB}$$

with the real number t varying with the point Z,

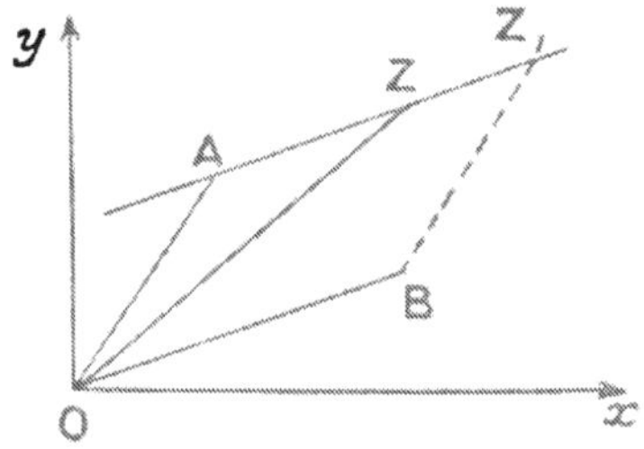

FIG. 24

$$\overline{OZ} = \overline{OA} + t\,\overline{OB}.$$

From this vector parametric equation of the line we obtain (7) equation (2). This result also follows (34) from the analytic parametric equations

$$x = a_1 + b_1 t, \qquad y = a_2 + b_2 t$$

of the line, where (a_1, a_2), (b_1, b_2) are the cartesian coordinates of the points A, B.

Corollaries. 1° *Every real line contains the point at infinity of the Gauss plane.* Equation (2) shows, in fact, that for t infinite we have z infinite.

2° It is easy to *calibrate the line* of equation (2), that is to say, to construct the points Z which correspond to diverse real values of t. Since the points A, Z_1 of affixes a, $a + b$ are given by the values 0,1 of t, the calibration is accomplished by considering AZ_1 as a positive unit segment.

3° *The parametric equation of the line joining the points* Z_1, Z_2 of affixes z_1, z_2 is

$$z = z_1 + (z_1 - z_2)t$$

since the vector $\overline{Z_2Z_1}$ having the direction of the line represents the number $z_1 - z_2$.

4° *To express that the points* Z_1, Z_2, Z_3 *are collinear*, it suffices to state that, t being a real number, we have

$$z_3 - z_1 = t(z_1 - z_2).$$

This equation is equivalent to

$$\overline{Z_1Z_3} = t\,\overline{Z_2Z_1}.$$

This way of expressing the collinearity of three points is generally preferred to that pointed out in article **37**.

37. Non-parametric equation. *The general equation of a real line of the Gauss plane is of the form*

$$\boxed{az + \bar{a}\bar{z} + b = 0} \qquad (3)$$

in which b *is a real number. This line contains the point of affix* — b/2a *and is perpendicular to the vector represented by the number* $\bar{a}$.

In fact, the general equation of a real line in rectangular cartesian coordinates is

$$\alpha x + \beta y + \gamma = 0,$$

the numbers α, β, γ being real. The real points Z of this line then have for affixes z the solutions of the equation (**33**)

$$\alpha (z + \bar{z}) + \frac{\beta}{i}(z - \bar{z}) + 2\gamma = 0$$

or

$$\alpha(z + \bar{z}) - i\beta(z - \bar{z}) + 2\gamma = 0,$$

$$(\alpha - i\beta)z + (\alpha + i\beta)\bar{z} + 2\gamma = 0$$

which, if one sets

$$\alpha - i\beta = a, \qquad \alpha + i\beta = \bar{a}, \qquad 2\gamma = b,$$

takes the form (3).

Equation (3) admits the solution $-b/2a$ inasmuch as b being real we have $\bar{b} = b$. The parallel line drawn through the origin has for equation

$$az + \bar{a}\bar{z} = 0,$$

and contains the point of affix $i\bar{a}$ and is thus perpendicular to the vector joining the origin to the point of affix $\bar{a}$.

Corollaries. 1° *The equation of the line joining the points* Z_1, Z_2 *of affixes* z_1, z_2 *is*

$$\begin{vmatrix} z & \bar{z} & 1 \\ z_1 & \bar{z}_1 & 1 \\ z_2 & \bar{z}_2 & 1 \end{vmatrix} = 0. \tag{4}$$

The line with equation (3) contains the points Z_1, Z_2 if we have

$$az_1 + \bar{a}\bar{z}_1 + b = 0, \tag{5}$$

$$az_2 + \bar{a}\bar{z}_2 + b = 0. \tag{6}$$

By claiming that the linear homogeneous equations (3), (5), (6) in a, $\bar{a}$, b are satisfied by values of a, $\bar{a}$, b which are not all zero, we obtain equation (4).

2° *The points* Z_1, Z_2, Z_3 *are collinear if*

$$\begin{vmatrix} z_1 & \bar{z}_1 & 1 \\ z_2 & \bar{z}_2 & 1 \\ z_3 & \bar{z}_3 & 1 \end{vmatrix} = 0.$$

3° *The line passing through the point of affix* z_1 *and parallel to the vector represented by the number* c *is*

$$\begin{vmatrix} z & \bar{z} & 1 \\ z_1 & \bar{z}_1 & 1 \\ c & \bar{c} & 0 \end{vmatrix} = 0.$$

In fact, the line contains the point of affix $z_1 + c$ and it suffices to refer to equation (4).

38. Centroid of a triangle. *If* a, b, c *are the affixes of the vertices* A, B, C *of a triangle* ABC, *which may be a degenerate triangle, then the affix* g *of the centroid* G *of the triangle is*

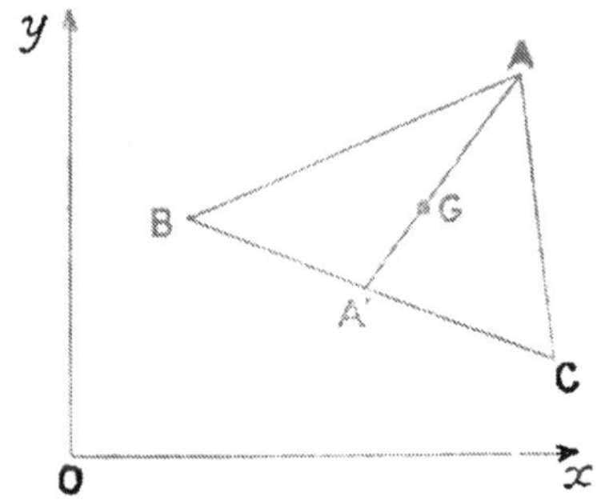

FIG. 25

$$g = \frac{1}{3}(a + b + c).$$

If A′ is the midpoint of segment BC, we have

$$a' = \frac{1}{2}(b + c)$$

and since

$$\frac{AG}{GA'} = 2,$$

we have (35)

$$g = \frac{a + 2a'}{1 + 2} = \frac{1}{3}(a + b + c).$$

39. Algebraic value of the area of a triangle. *If the angle* (xy) *has the value* $+\pi/2$ *and if in tracing the perimeter of a triangle one meets the vertices* A, B, C *of affixes* a, b, c *in this order, then the algebraic value* ABC *of the area of the triangle is*

$$\mathrm{ABC} = \frac{i}{4}\begin{vmatrix} a & \bar{a} & 1 \\ b & \bar{b} & 1 \\ c & \bar{c} & 1 \end{vmatrix}.$$

The algebraic value is, in fact, that of $(\overline{AB} \times \overline{AC})/2$ on an axis $O\zeta$ such that the trihedral $Oxy\zeta$ is trirectangular and right-handed. That is to say (**11**)

$$\frac{i}{4}[(b - a)(\bar{c} - \bar{a}) - (\bar{b} - \bar{a})(c - a)] =$$

$$\frac{i}{4}\begin{vmatrix} b - a & \bar{b} - \bar{a} \\ c - a & \bar{c} - \bar{a} \end{vmatrix} = \frac{i}{4}\begin{vmatrix} a & \bar{a} & 1 \\ b - a & \bar{b} - \bar{a} & 0 \\ c - a & \bar{c} - \bar{a} & 0 \end{vmatrix} = \frac{i}{4}\begin{vmatrix} a & \bar{a} & 1 \\ b & \bar{b} & 1 \\ c & \bar{c} & 1 \end{vmatrix}.$$

Corollaries. 1° *The algebraic values*

ABC, BCA, CAB

are equal, and they differ only in sign from the equal values

ACB, CBA, BAC.

This follows since a determinant merely changes sign if two rows are exchanged.

2° *If* P *is any point whatever of the plane, then*

$$\mathrm{PBC} + \mathrm{PCA} + \mathrm{PAB} = \mathrm{ABC}, \qquad (7)$$

$$\overline{\mathrm{PA}}.\mathrm{PBC} + \overline{\mathrm{PB}}.\mathrm{PCA} + \overline{\mathrm{PC}}.\mathrm{PAB} = 0. \qquad (8)$$

By taking the origin of cartesian axes at the point P, we see that
PBC + PCA + PAB =

$$\frac{i}{4}\left[\begin{vmatrix} 0 & 0 & 1 \\ b & \bar{b} & 1 \\ c & \bar{c} & 1 \end{vmatrix} + \begin{vmatrix} 0 & 0 & 1 \\ c & \bar{c} & 1 \\ a & \bar{a} & 1 \end{vmatrix} + \begin{vmatrix} 0 & 0 & 1 \\ a & \bar{a} & 1 \\ b & \bar{b} & 1 \end{vmatrix}\right] = \frac{i}{4}\begin{vmatrix} a & \bar{a} & 1 \\ b & \bar{b} & 1 \\ c & \bar{c} & 1 \end{vmatrix} = \mathrm{ABC}.$$

Furthermore
$a\,\mathrm{PBC} + b\,\mathrm{PCA} + c\,\mathrm{PAB} =$

$$\frac{i}{4}\left[a\,(b\bar{c} - \bar{b}c) + b\,(c\bar{a} - \bar{c}a) + c\,(a\bar{b} - \bar{a}b)\right] = 0$$

and this equation establishes (7) equation (8).

3° *If* p *is the affix of any point* P *whatever in the plane of a non-degenerate triangle whose vertices* A, B, C *have* a, b, c *for affixes, then*

$$p = a\frac{\mathrm{PBC}}{\mathrm{ABC}} + b\frac{\mathrm{PCA}}{\mathrm{ABC}} + c\frac{\mathrm{PAB}}{\mathrm{ABC}}. \qquad (9)$$

In fact, equation (8) is equivalent to

$$(a - p)\mathrm{PBC} + (b - p)\mathrm{PCA} + (c - p)\mathrm{PAB} = 0,$$

from which we obtain (9).

The numbers PBC/ABC, PCA/ABC, PAB/ABC are the absolute barycentric coordinates of P for triangle ABC. *The relation* (9) *thus expresses the affix of a point as a function of the absolute barycentric coordinates of this point for any triangle and of the affixes of the vertices of the triangle.*

Exercises 32 through 37

32. Derive the non-parametric equation of a straight line from the parametric equation, and vice versa. [If $z = a + bt$, we can refer to corollary 3° of article **37**, or better, by a frequently used general idea, adjoin the conjugate equation

$$\bar{z} = \bar{a} + \bar{b}t$$

and eliminate t. Vice versa, for

$$az + \overline{az} + b = 0,$$

we know a point of the line and its direction.]

33. Intersection of two lines d_1, d_2. 1° Find the point of intersection of the two lines given by

$$a_1 z + \bar{a}_1 \bar{z} + b_1 = 0, \qquad a_2 z + \bar{a}_2 \bar{z} + b_2 = 0, \qquad b_1, b_2 \text{ real.}$$

[If

$$a_1 \bar{a}_2 - \bar{a}_1 a_2 \neq 0,$$

the point of intersection is

$$z = \frac{\bar{a}_1 b_2 - \bar{a}_2 b_1}{a_1 \bar{a}_2 - a_2 \bar{a}_1}.$$

If

$$a_1/a_2 = \bar{a}_1/\bar{a}_2,$$

the lines are parallel or coincide according as these ratios are or are not different from b_1/b_2.]

2° Find the point of intersection of the two lines given by

$$z = a_1 + b_1 t_1, \qquad z = a_2 + b_2 t_2, \qquad t_1, t_2 \text{ real.}$$

[Reduce to 1° (exercise 32), or adjoin

$$\bar{a}_1 + \bar{b}_1 t_1 = \bar{a}_2 + \bar{b}_2 t_2 \qquad \text{to} \qquad a_1 + b_1 t_1 = a_2 + b_2 t_2,$$

find t_1 if possible, ...]

34. Concurrent lines. Find a necessary and sufficient condition for the lines

$$a_i z + \bar{a}_i \bar{z} + b_i = 0 \qquad \text{or} \qquad z = a_i + b_i t_i, \qquad i = 1, 2, 3$$

to pass through a common point. [Answer:

$$\begin{vmatrix} a_1 & \bar{a}_1 & b_1 \\ a_2 & \bar{a}_2 & b_2 \\ a_3 & \bar{a}_3 & b_3 \end{vmatrix} = 0 \qquad \text{or} \qquad \begin{vmatrix} b_1 & \bar{b}_1 & a_1\bar{b}_1 - \bar{a}_1 b_1 \\ b_2 & \bar{b}_2 & a_2\bar{b}_2 - \bar{a}_2 b_2 \\ b_3 & \bar{b}_3 & a_3\bar{b}_3 - \bar{a}_3 b_3 \end{vmatrix} = 0.]$$

35. Perpendicularity. 1° The lines having equations

$$a_1 z + \bar{a}_1 \bar{z} + b_1 = 0, \qquad a_2 z + \bar{a}_2 \bar{z} + b_2 = 0, \qquad b_1, b_2 \text{ real}$$

are perpendicular if

$$a_1 \bar{a}_2 + a_2 \bar{a}_1 = 0.$$

[See articles **11** and **37**.]

2° The equation of the perpendicular to the first line and passing through the point of affix z_0 is

$$a_1(z - z_0) - \bar{a}_1(\bar{z} - \bar{z}_0) = 0.$$

Treat the case where the line is given parametrically. In each case, calculate the affix of the foot of the perpendicular.

36. The perpendicular bisector of side BC of a triangle ABC has for equation

$$z(\bar{b} - \bar{c}) + \bar{z}(b - c) = b\bar{b} - c\bar{c}.$$

The three perpendicular bisectors are concurrent at the center O of the circumscribed circle with affix

$$-\begin{vmatrix} a\bar{a} & a & 1 \\ b\bar{b} & b & 1 \\ c\bar{c} & c & 1 \end{vmatrix} : \begin{vmatrix} a & \bar{a} & 1 \\ b & \bar{b} & 1 \\ c & \bar{c} & 1 \end{vmatrix}.$$

The altitude through vertex A has for equation

$$z(\bar{b} - \bar{c}) + \bar{z}(b - c) = a(\bar{b} - \bar{c}) + \bar{a}(b - c).$$

The three altitudes are concurrent at the orthocenter H with affix

$$[\bar{a}(c - b)(c + b - a) + \bar{b}(a - c)(a + c - b) + \bar{c}(b - a)(b + a - c)] : \begin{vmatrix} a & \bar{a} & 1 \\ b & \bar{b} & 1 \\ c & \bar{c} & 1 \end{vmatrix}.$$

The median through vertex A has for equation

$$z(2\bar{a} - \bar{b} - \bar{c}) - \bar{z}(2a - b - c) = \bar{a}(b + c) - a(\bar{b} + \bar{c}).$$

We know (article 38) that the point of concurrency G of the medians has affix

$$(a + b + c)/3.$$

More generally, we designate as the centroid of n points Z_i the point with affix $(\Sigma z_i)/n$.

37. Determine the algebraic value of the area of the triangle formed by the lines having equations

$$a_i z + \bar{a}_i \bar{z} + b_i = 0, \qquad i = 1, 2, 3.$$

[See article 39 and exercise 33.]

III. THE CIRCLE

40. Non-parametric equation. *The general equation of a real or ideal circle of the Gauss plane is of the form*

$$\boxed{z\bar{z} + az + \bar{a}\bar{z} + b = 0} \qquad (1)$$

in which b *is a real number.*

The affix of the center is $-\bar{a}$ *and the square of the radius is* $a\bar{a} - b$.

In rectangular cartesian coordinates, every real or ideal circle has an equation of the form

$$x^2 + y^2 + 2\alpha x + 2\beta y + \gamma = 0,$$

the numbers α, β, γ being real. The coordinates of the center are $-\alpha$, $-\beta$ and the square of the radius is $\alpha^2 + \beta^2 - \gamma$. According as this square is or is not positive, the circle is real or ideal.

In complex coordinates, the equation of the circle is then (33)

$$z\bar{z} + \alpha(z + \bar{z}) - i\beta(z - \bar{z}) + \gamma = 0$$

or

$$z\bar{z} + (\alpha - i\beta)z + (\alpha + i\beta)\bar{z} + \gamma = 0$$

which, if we set

$$\alpha - i\beta = a, \quad \alpha + i\beta = \bar{a}, \quad \gamma = b, \qquad (2)$$

takes the form (1).

From equations (2) we find that the affix of the center and the square of the radius are

$$-\alpha - i\beta = -\bar{a}, \quad \alpha^2 + \beta^2 - \gamma = (\alpha + i\beta)(\alpha - i\beta) - \gamma = a\bar{a} - b.$$

Example. *If* A, B, C *are three given points in the Gauss plane, let us determine the locus of points* Z *such that*

$$\overline{ZA}.\overline{ZB} + \overline{ZB}.\overline{ZC} + \overline{ZC}.\overline{ZA} = 0.$$

If a, b, c, z are the affixes of A, B, C, Z, we must have (**10**)

$$(a - z)(\bar{b} - \bar{z}) + (\bar{a} - \bar{z})(b - z) + (b - z)(\bar{c} - \bar{z}) + (\bar{b} - \bar{z})(c - z) + (c - z)(\bar{a} - \bar{z}) + (\bar{c} - \bar{z})(a - z) = 0$$

or, by expanding,

$$6 z\bar{z} - 2(\bar{a} + \bar{b} + \bar{c}) z - 2(a + b + c)\bar{z} + (a + b + c)(\bar{a} + \bar{b} + \bar{c}) - (a\bar{a} + b\bar{b} + c\bar{c}) = 0.$$

The locus is thus a circle. If we place the origin at the centroid G of the system of points A, B, C, we have (**38**)

$$a + b + c = \bar{a} + \bar{b} + \bar{c} = 0,$$

$$a\bar{a} = GA^2 = \frac{1}{9}(2\,AB^2 + 2\,AC^2 - BC^2)$$

and the equation of the circle becomes

$$z\bar{z} = \frac{1}{6}(GA^2 + GB^2 + GC^2) = \frac{1}{18}(AB^2 + BC^2 + CA^2).$$

The center is G and the square of the radius is the value of $z\bar{z}$.[1]

41. Parametric equation. *In the Gauss plane, the equation*

$$\boxed{z = \frac{at + b}{ct + d}} \tag{3}$$

in which a, b, c, d *are real or imaginary constants such that*

$$ad - bc \neq 0 \tag{4}$$

and t *is a parameter able to take on all real values, represents*

1° *a straight line if* c *is zero or if* d/c *is real;*

2° *a circle in all other cases.*

[1] See exercise 224 of *Compléments de géométrie.*

When $c = 0$, we have $ad \neq 0$ by virtue of (4), and equation (3), which now becomes

$$z = \frac{a}{d}t + \frac{b}{d}$$

represents a straight line (**36**).

When $d = 0$, we have $bc \neq 0$, and by setting

$$\frac{1}{t} = \mathrm{T}$$

equation (3) becomes

$$z = \frac{b}{c}\mathrm{T} + \frac{a}{c}.$$

This represents a straight line.

Now suppose $cd \neq 0$. To the values t, 1, 0, ∞ of the parameter t there correspond on the sought locus the points Z, Z_1, Z_0, Z_∞ of affixes

$$z, \quad z_1 = \frac{a+b}{c+d}, \quad z_0 = \frac{b}{d}, \quad z_\infty = \frac{a}{c}.$$

The points Z_1, Z_0, Z_∞ are distinct, for we have

$$z_1 - z_0 = \frac{ad - bc}{d(c+d)}, \quad z_1 - z_\infty = \frac{bc - ad}{c(c+d)}, \quad z_0 - z_\infty = \frac{bc - ad}{cd}$$

as well as relation (4). Furthermore

$$(z\, z_1 z_0 z_\infty) = \frac{z - z_0}{z_1 - z_0} : \frac{z - z_\infty}{z_1 - z_\infty} = \frac{\dfrac{(ad - bc)\,t}{d(ct+d)}}{\dfrac{ad - bc}{d(c+d)}} : \frac{\dfrac{bc - ad}{c(ct+d)}}{\dfrac{bc - ad}{c(c+d)}} = t.$$

The four points Z, Z_1, Z_0, Z_∞ are therefore on a straight line or on a circle (**28**). The curve, which is determined by the fixed points Z_1, Z_0, Z_∞, is the locus of all the points Z furnished by the set of real values of t considered in (3). The curve is a straight line if there exists a real value of t which makes z infinite, that is to say, if the root $-d/c$ of

$$ct + d = 0$$

is real, for a straight line is distinguished from a circle by the fact that it contains the point at infinity of the Gauss plane.

Conversely, any straight line and any real circle can be represented by an equation of the form (3).

In fact, if Z, P, Q, R are an arbitrary point and three fixed points, of affixes z, p, q, r, on a straight line or circle, we have (28), t denoting a real number varying with Z,

$$(zpqr) = t$$

or, after the development of

$$\frac{z-q}{p-q} : \frac{z-r}{p-r} = t,$$

$$z = \frac{r(p-q)t - q(p-r)}{(p-q)t - (p-r)}.$$

In this equation, which has the form (3), we have

$$a = r(p-q), \quad b = q(r-p), \quad c = p-q, \quad d = r-p,$$
$$ad - bc = (p-q)(r-p)(r-q)$$

and this product is not zero since P, Q, R are distinct.

42. Construction and calibration. To construct a circle given by an equation of the form (3), it suffices to construct three points of the circle. It is particularly interesting to construct the points Z_1, Z_0, Z_∞ corresponding to the values 1, 0, ∞ of t and whose affixes are

$$z_1 = \frac{a+b}{c+d}, \quad z_0 = \frac{b}{d}, \quad z_\infty = \frac{a}{c}.$$

But if these points are close to one another, it is expedient to use other values of t. Or we can calculate, as follows, the affix ω of the center and the magnitude R of the radius. It suffices to find the non-parametric equation of the circle by eliminating t from equation (3) and its conjugate

$$t(\bar{c}\bar{z} - \bar{a}) = \bar{b} - \bar{d}\bar{z},$$

which gives

$$z\bar{z} - \frac{\bar{a}d - \bar{b}c}{\bar{c}d - \bar{d}c} z - \frac{a\bar{d} - b\bar{c}}{c\bar{d} - d\bar{c}} \bar{z} + \frac{a\bar{b} - b\bar{a}}{c\bar{d} - d\bar{c}} = 0.$$

From this we obtain (40)

$$\omega = \frac{a\bar{d} - b\bar{c}}{c\bar{d} - d\bar{c}},$$

$$R = |\omega - z_0| = \left| \frac{a\bar{d} - b\bar{c}}{c\bar{d} - d\bar{c}} - \frac{b}{d} \right| = \left| \frac{\bar{d}(ad - bc)}{d(c\bar{d} - d\bar{c})} \right| = \left| \frac{ad - bc}{c\bar{d} - d\bar{c}} \right|.$$

We shall obtain these results again by other methods in articles **61** and **127**.

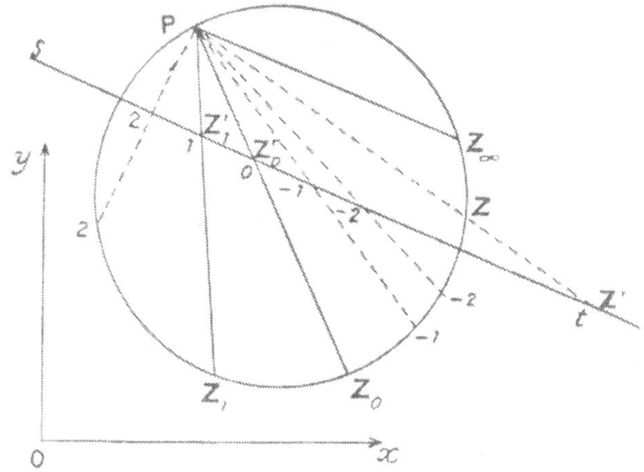

FIG. 26

To calibrate the circle, take any point P on it and cut the pencil $P(ZZ_1Z_0Z_\infty)$ by any secant s parallel to the ray PZ_∞. Then we have (41)

$$t = (ZZ_1Z_0Z_\infty) = P(ZZ_1Z_0Z_\infty) = (Z'Z'_1Z'_0\infty),$$

$$t = \frac{Z'_0Z'}{Z'_0Z'_1}.$$

Hence, if we adopt as the positive sense of s that from Z'_0 to Z'_1 and for unity the length of the segment $Z'_0Z'_1$, we have $Z'_0Z' = t$. The s-axis is immediately calibrated, and it suffices to project this calibration from P onto the circle.

43. Particular cases. 1° *Let* A, B *be two given points of affixes* a, b *and let* θ *be a given real number. The arc described on* AB *and for which* AB *subtends an angle of algebraic value* θ *has for equation*

$$\frac{z - b}{z - a} = te^{i\theta}, \qquad (5)$$

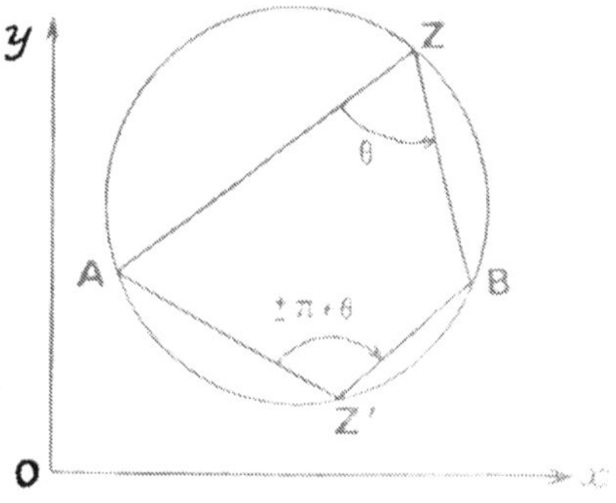

FIG. 27

t *being able to assume all real non-negative values.*

For t < 0, *this equation represents the arc described on* AB *and for which* AB *subtends an angle of algebraic value* $\pm \pi + \theta$.

If Z is any point of the first arc, we have

$$(\overline{ZA}, \overline{ZB}) = \theta$$

and (17)

$$\frac{z-b}{z-a} = \left|\frac{ZB}{ZA}\right| e^{i\theta} = te^{i\theta}.$$

For a point Z′ of the remaining portion of the circumference, we have

$$(\overline{Z'A}, \overline{Z'B}) = \pm\pi + \theta, \quad \frac{z'-b}{z'-a} = \left|\frac{Z'B}{Z'A}\right| e^{i(\pm\pi+\theta)} = -t'e^{i\theta},$$

$$t' \geqslant 0.$$

Equation (5) is of the form (3) since it can be written as

$$z = \frac{ate^{i\theta} - b}{te^{i\theta} - 1}.$$

2° *The circle which is the locus of points* Z *the ratio of whose distances from two given points* A, B *is a given positive number* k *different from* 1 *has for equation*

$$\frac{z-a}{z-b} = ke^{it}, \tag{6}$$

z, a, b *being the affixes of* Z, A, B *and* t *being a real parameter.*

If k = 1, *equation* (6) *is that of the perpendicular bisector of segment* AB.

Notice that equation (6) takes the form (3) if we set

$$\tan\frac{t}{2} = t_1$$

inasmuch as

$$e^{it} = \cos t + i \sin t = \frac{1 - t_1^2 + 2it_1}{1 + t_1^2} = \frac{(1 + it_1)^2}{(1 + it_1)(1 - it_1)} = \frac{1 + it_1}{1 - it_1}$$

and (6) becomes

$$z = \frac{i(kb + a)\, t_1 + kb - a}{i(k + 1)\, t_1 + k - 1}.$$

Corollary. *If* u, v *are two real parameters, then in the parametric representation*

$$\frac{z-a}{z-b} = ue^{iv}$$

of the Gauss plane, the parametric curves v = constant *are the circles of the pencil having* A, B *for base points, and the curves* u = constant *are the circles of the orthogonal pencil.*

44. Cases where ad — bc = 0. Let us find the points Z whose affixes are roots of equation (3) written in the form

$$z(ct + d) = at + b.$$

1st case, $c = d = 0$. If $a = b = 0$, every point of the plane is a point Z. If $a = 0$ and $b \neq 0$, or if $a \neq 0$ and b/a is complex imaginary, there is no point Z. If $a \neq 0$ and b/a is real, the value $t = -b/a$ yields all the points of the plane for points Z.

2nd case, $c = 0$, $d \neq 0$. We have $a = 0$ and the single point Z of affix b/d.

3rd case, $c \neq 0$. We have $b = ad/c$ and the equation becomes

$$\left(z - \frac{a}{c}\right)(ct + d) = 0.$$

If d/c is complex imaginary, the single point Z has a/c for affix. If d/c is real, we have this same point for $t \neq -d/c$ and every point of the plane for $t = -d/c$.

45. Example. *Being given three non-collinear points* A, B, C *and a real number* θ, *let us determine the points* Z *such that if we rotate the points* B, C *about* Z *through angles of algebraic values* 2θ *and* θ *respectively, the resulting points* B′, C′ *shall be collinear with point* A.

Consider the figure in the oriented Gauss plane and related to two perpendicular axes Ox, Oy for which $(xy) = +\pi/2$. If a, b, c, z, b', c' are the affixes of points A, B, C, Z, B′, C′, we have (15)

$$b' - z = (b - z)\, e^{2i\theta}, \quad (7)$$

$$c' - z = (c - z)\, e^{i\theta}, \quad (8)$$

and the points B′, C′, A are collinear if (36), t being a real number, we have

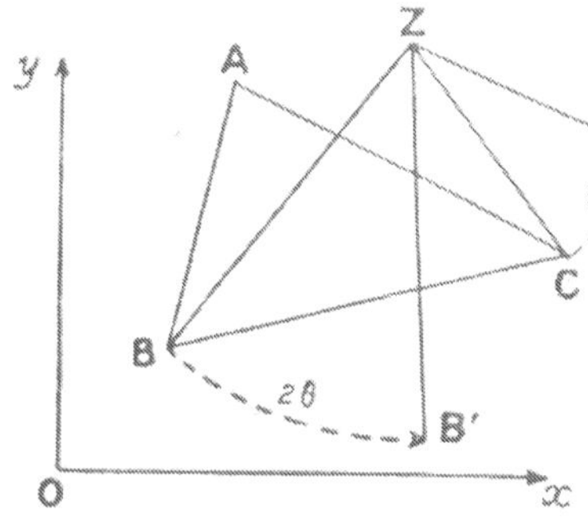

FIG. 28

$$b' - a = t(c' - a).$$

Upon replacing b' and c' by their values obtained from (7), (8), this equation becomes

$$z(1 - e^{2i\theta}) + be^{2i\theta} - a = t[z(1 - e^{i\theta}) + ce^{i\theta} - a]$$

or

$$z[(1 - e^{i\theta})t - (1 - e^{2i\theta})] = (a - ce^{i\theta})t - (a - be^{2i\theta}) \qquad (9)$$

and is of the form (**41**)

$$z(ct + d) = at + b.$$

The expression $ad - bc$ for equation (9) is

$$e^{i\theta}(e^{i\theta} - 1)\,[a - c + (b - c)e^{i\theta}]$$

and is zero for either one of the following two cases,

$$e^{i\theta} = 1 \quad \text{or} \quad \theta = 2\,n\pi,\ n \text{ an integer}; \quad \frac{a - c}{c - b} = e^{i\theta}.$$

1st case, $\theta = 2n\pi$. Equation (9) becomes

$$oz = (a - c)\,t - (a - b).$$

The value $(a - b)/(a - c)$ for t, which makes the second member vanish, is not real since A, B, C are not collinear, and hence the equation has no root and *there is no point* Z. Besides, the rotations give B′ = B, C′ = C, which explains the non-existence of Z.

2nd case. The equation

$$\frac{a - c}{c - b} = e^{i\theta} \quad \text{or} \quad \frac{c - a}{c - b} = e^{i(\pi+\theta)} \qquad (10)$$

says that $|\,CA\,| = |\,CB\,|$ and that $(\overline{CB}, \overline{CA}) = \pi + \theta$ to within an integral multiple of 2π.

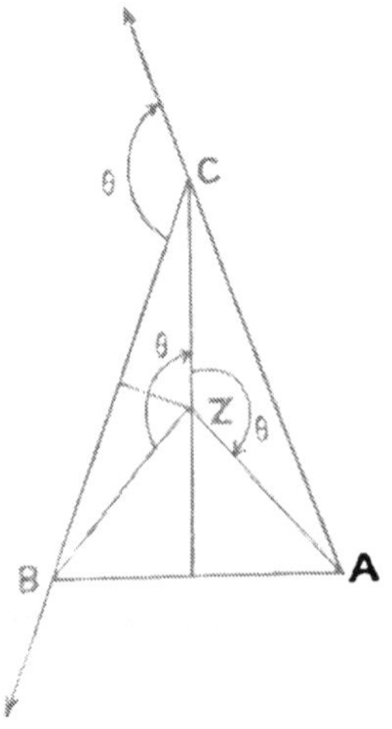

FIG. 29

If we take note of the value (10) of $e^{i\theta}$, equation (9) becomes

$$[z(2c - a - b) - (c^2 - ab)]\,[t(c - b) - (a - b)] = 0.$$

Since the second factor cannot vanish inasmuch as $(a - b)/(c - b)$ is complex imaginary, *there is a unique point* Z and its affix is

$$z = \frac{c^2 - ab}{2c - a - b}. \qquad (11)$$

To obtain a simple construction of Z, write (11) successively as

$$c^2 - 2cz = -z\,(a + b) + ab,$$
$$c^2 - 2cz + z^2 = z^2 - z(a + b) + ab,$$
$$(z - c)^2 = (z - a)\,(z - b).$$

The point Z is then (**31**, 4°) the midpoint of the segment which joins C to its harmonic conjugate C_1 with respect to points A and B. Since triangle CAB is isosceles, the symmedian CC_1 is a diameter of the circumscribed circle, and point Z is the center of the circumscribed circle of the triangle. We easily recognize the fact that the rotations 2θ and θ about Z carry B, C into $B' = A = C'$.

We are led more directly to the indicated construction by the following often employed device: we choose the origin of the Ox and Oy axes in such a fashion as to simplify the affix of the point to be constructed (see the example of article **40**). If we place the origin at C, we have $c = 0$ and equation (11) becomes

$$z = \frac{ab}{a+b} \quad \text{or} \quad \frac{1}{z} = \frac{1}{a} + \frac{1}{b}.$$

The construction of Z follows from (**31**, 2°).

Suppose now that $ad - bc \neq 0$. There is then a locus (a straight line or a circle) of points Z.

3rd case. The coefficient of z in (9) being

$$(1 - e^{i\theta})(t - 1 - e^{i\theta}),$$

the locus is a straight line if $1 + e^{i\theta}$ is real, that is, if $e^{i\theta}$ has the value 1 or -1. The first hypothesis has to be excluded (1st case). For $e^{i\theta} = -1$, θ is an odd multiple of π and the equation is

$$z = \frac{t(a + c) - (a - b)}{2t}$$

or, setting $1/2t = t_1$,

$$z = (b - a)\, t_1 + \frac{a + c}{2}.$$

The sought line is then the parallel to line AB drawn through the midpoint of segment AC. Moreover, we immediately see that for the rotations under consideration C′ is on AB and $B' = B$.

4th case. *The locus is a circle* in all other cases. The points Z_0, Z_∞, Z_1 given by the values 0, ∞, 1 of t have for affixes

$$z_0 = \frac{a - be^{2i\theta}}{1 - e^{2i\theta}}, \quad z_\infty = \frac{a - ce^{i\theta}}{1 - e^{i\theta}}, \quad z_1 = \frac{c - be^{i\theta}}{1 - e^{i\theta}}.$$

From these we obtain

$$\frac{z_0 - a}{z_0 - b} = e^{2i\theta}, \quad \frac{z_\infty - a}{z_\infty - c} = e^{i\theta}, \quad \frac{z_1 - c}{z_1 - b} = e^{i\theta}.$$

The point Z_0 is then the intersection of the perpendicular bisector m_c of segment AB with the arc γ described on AB and for which AB subtends an angle of algebraic value $(\overline{Z_0B}, \overline{Z_0A}) = 2\theta$. We have similar constructions for Z_∞ and Z_1 with the couples (m_b, β) and (m_a, α). The required circle l passes through Z_0, Z_∞, Z_1.

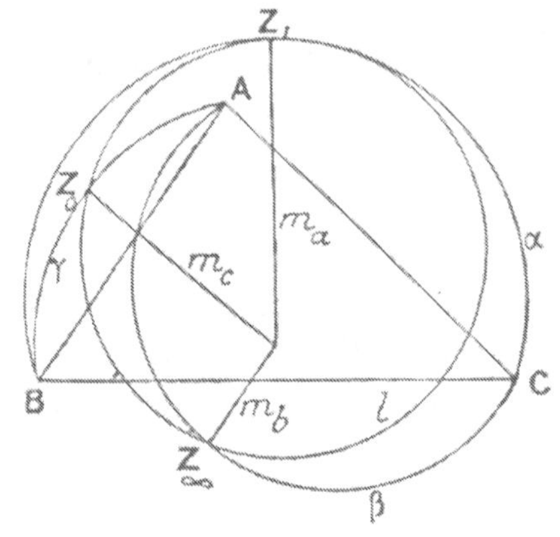

FIG. 30

We easily recognize the fact that Z_0, Z_∞, Z_1 must belong to the locus. The rotation $(Z_0, 2\theta)$ carries B into A; (Z_∞, θ) gives $C' = A$; $(Z_1, 2\theta)$ and (Z_1, θ) give $B' = C'$.

Exercises 38 through 45

38. Construct on quadrilruled paper the locus having parametric equation

$$z = \frac{1 + it}{1 - it}.$$

Calibrate this locus in such a way that its points $t = 0$, $t = 1$ will also be the points $t = 0$, $t = 1$ of the auxiliary linear scale. Show that two points of the locus which correspond to the values t and $-t$ are symmetric with respect to the Ox axis, and that the given equation can also be written as $z = e^{it_1}$, where t_1 is a real parameter.

39. The locus with equation

$$z = \frac{(0.22 + 0.46i)t - 1}{(0.2 + 0.6i)t - 1}$$

is a circle whose center, radius, and calibration on quadrilruled paper is required. [Answer: $53/60 - i/60$; $\sqrt{50}/60$.]

40. Choice of a unit circle. When the properties of a figure to be studied depend essentially upon three non-collinear points A, B, C, it is advantageous, for simplicity of calculation, to choose for unity the length of the radius of the circle (O) circumscribed about triangle ABC and to place the center O of this circle at the origin of the rectangular axes Ox, Oy of the Gauss plane. Circle (O) is then referred to as the unit circle. We can still choose the position of Ox, for example by taking as real the affix of some point playing an important role.

Since the equation of (O) is $z\bar{z} = 1$, we have

$$\bar{a} = \frac{1}{a}, \qquad \bar{b} = \frac{1}{b}, \qquad \bar{c} = \frac{1}{c}.$$

If we set

$$s_1 = a + b + c, \qquad s_2 = bc + ca + ab, \qquad s_3 = abc,$$

a, b, c are the roots of the cubic equation

$$z^3 - s_1 z^2 + s_2 z - s_3 = 0,$$

and we have the relations of exercise 11. The point with affix s_2/s_1 is on (O).

41. With the axes of exercise 40, show that in a triangle ABC :

1° the equation of the line BC is (article **37**, corollary 1°)

$$z + bc\bar{z} = b + c;$$

2° that of the altitude AA_1 is (exercise 35)

$$z - bc\bar{z} = a - \frac{bc}{a};$$

3° the affix of the foot A_1 of this altitude is

$$a_1 = \tfrac{1}{2}\left(s_1 - \frac{s_3}{a^2}\right);$$

4° the three altitudes AA_1, BB_1, CC_1 are concurrent at a point H (the orthocenter) with affix s_1, situated on the *Euler line* joining the circumcenter O to the centroid G of the triangle, and such that $\overline{OH} = 3\overline{OG}$;

5° the centroid of the four points A, B, C, H is the midpoint O_9 of OH;

6° the algebraic value of the area of the *orthic triangle* $A_1B_1C_1$ is

$$\frac{-i(b^2 - c^2)(c^2 - a^2)(a^2 - b^2)}{16a^2b^2c^2};$$

7° the tangents to circle (O) at the points A, B, C form a triangle A′B′C′ (the *tangential triangle*) whose vertex A′ has affix [because (OA′BC) = —1]

$$a' = \frac{2bc}{b + c}.$$

and whose area has the algebraic value

$$\frac{-i(b - c)(c - a)(a - b)}{(b + c)(c + a)(a + b)};$$

8° the symmedian AA′ has for equation

$$z(3a - s_1) + \bar{z}(as_2 - 3s_3) + 2(bc - a^2) = 0,$$

and the three symmedians are concurrent at the *Lemoine point* K with affix

$$k = \frac{2(3s_1 - s_2\bar{s}_1)}{9 - s_1\bar{s}_1} = \frac{2(3s_1s_3 - s_2^2)}{9s_3 - s_1s_2};$$

9° the isodynamic centers W, W′ are given by the equation (exercise 27)

$$(3s_2 - s_1^2)z^2 + (s_1s_2 - 9s_3)z + 3s_1s_3 - s_2^2 = 0$$

which, since it can be written as (8°)

$$\bar{k}z^2 - 2z + k = 0,$$

proves that W, W' are collinear with the points O, K, which they harmonically separate, and are inverses of one another in circle (O). [Place Ox on OK.]

42. Second parametric representation of a circle. A point Z moving on a circle of center Ω and radius R has an affix of the form

$$z = \omega + \mathrm{R}e^{it}, \qquad t \text{ a real variable,}$$

in which we frequently designate the exponential e^{it} by τ.

Show that *the equation*

$$z = \frac{a\tau + b}{c\tau + d}, \qquad ad - bc \neq 0, \qquad \tau = e^{it}, \qquad t \text{ real}$$

represents a straight line when $|c| = |d|$; *otherwise it represents a circle whose center* Ω *and radius* R *are given by*

$$\omega = \frac{a\bar{c} - b\bar{d}}{c\bar{c} - d\bar{d}}, \qquad \mathrm{R} = \left| \frac{ad - bc}{c\bar{c} - d\bar{d}} \right|.$$

[As in 2° of article **43**, set tan $t/2 = t_1$, which reduces the equation to

$$z = \frac{a + b + i(a - b)t_1}{c + d + i(c - d)t_1},$$

and use the results of article **41**. The development is shorter if we use articles **76** and **91**.]

43. The nine-point or Feuerbach (or Euler) circle. In a triangle ABC, the feet A_1, B_1, C_1 of the altitudes, the midpoints A_2, B_2, C_2 of the sides, and the midpoints A_3, B_3, C_3 of the distances from A, B, C to the orthocenter H are on a circle (O_9) whose center O_9 is the midpoint of OH and whose radius is half that of the circumscribed circle. [We have (exercise 41, 3°)

$$a_1 = \frac{s_1}{2} - \frac{s_3}{2a^2},$$

whence A_1, B_1, C_1 are on the circle (exercise 42) having equation

$$z = \frac{s_1}{2} - \left(\frac{s_3}{2}\right)\tau,$$

center $s_1/2$, and radius $|s_3/2| = 1/2$. Similarly for a_2 and a_3.]

44. If R is the (unit) radius of the circle (O) circumscribed about triangle ABC, the circumscribed circle of the tangential triangle A'B'C' (exercise 41, 7°) is the inverse of the Feuerbach circle in the inversion of center O and power R^2 [since $\mathrm{OA}_2 \cdot \mathrm{OA}' = \mathrm{R}^2$ if A_2 is the midpoint of BC]. Show (article 19) that the equation of circle (A'B'C') is (exercise 43)

$$z = \frac{2}{\bar{s}_1 - \bar{s}_3\tau},$$

whose center M has affix

$$\frac{2s_1}{s_1\bar{s}_1 - 1}$$

and lies on the Euler line such that

$$\mathrm{OM} = \frac{2\mathrm{R}^2 \cdot \mathrm{OH}}{\mathrm{OH}^2 - \mathrm{R}^2};$$

the radius of the circle is

$$\frac{2\mathrm{R}^3}{|\,\mathrm{OH}^2 - \mathrm{R}^2\,|}.$$

45. Griffiths' pencil. Show that the equation of the pencil of circles determined by the circumcircle (O) and the Feuerbach circle (O_9) is

$$4(1 - \lambda)z\bar{z} - 2\bar{s}_1 z - 2s_1\bar{z} + 4\lambda + s_1\bar{s}_1 - 1 = 0,$$

where λ is a real parameter.

The equation of the radical axis of the pencil is

$$2\bar{s}_1 z + 2s_1\bar{z} = 3 - s_1\bar{s}_1.$$

The affix of the center and the square of the radius of a circle of the pencil are

$$\frac{s_1}{2(1 - \lambda)}, \qquad \frac{\lambda s_1\bar{s}_1 - (4\lambda - 1)(1 - \lambda)}{4(1 - \lambda)^2}.$$

For

$$\lambda = -\frac{1}{2}, \quad \frac{1}{2}, \quad \frac{1}{4}, \quad \frac{5 - s_1\bar{s}_1}{4}$$

we have the orthoptic circle, with center G, of the Steiner ellipse inscribed in triangle ABC, the conjugate circle, with center H, of the triangle, the orthocentroidal circle, with diameter HG, of the triangle, and the circumcircle of the tangential triangle (exercise 44).

IV. THE ELLIPSE

46. Generation with the aid of two rotating vectors.

If two vectors issued from a fixed point O *have given but different magnitudes, and if they rotate about* O *with constant and opposite angular velocities* ω *and* $-\omega$, *the fourth vertex of the parallelogram having the two vectors for a pair of adjacent sides describes an ellipse* (E) *of center* O.

If, for a rectangular cartesian system of origin O, *the affixes of the extremities* A, B *of the vectors, in some one of their positions, are* a, b, *then the parametric equation of* (E) *is*

$$\boxed{z = ae^{i\omega t} + be^{-i\omega t},} \tag{1}$$

the parameter t *being allowed to take on all real values.*

Let $\overline{\text{OA}}_t$, $\overline{\text{OB}}_t$ be the positions assumed by the vectors $\overline{\text{OA}}$, $\overline{\text{OB}}$ at the end of time t, and let Z be the fourth vertex of the parallelogram constructed on OA_t, OB_t; denote the affixes of A_t, B_t, Z by a_t, b_t, z. We have (15)

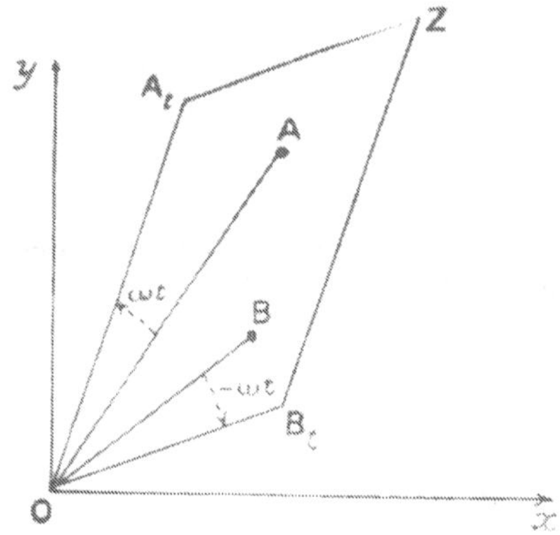

FIG. 31

$$a_t = ae^{i\omega t}, \quad b_t = be^{-i\omega t}$$

and, since

$$\overline{\text{OZ}} = \overline{\text{OA}}_t + \overline{\text{OB}}_t,$$

we obtain the equation (7)

$$z = a_t + b_t$$

or (1) for the locus described by Z.

To find the ordinary cartesian equation of this locus we can write (1) in the form

$$x + iy = (a_1 + ia_2)(\cos \omega t + i \sin \omega t) + (b_1 + ib_2)(\cos \omega t - i \sin \omega t),$$

separate the real and the imaginary parts, then eliminate t from the two parametric equations so obtained.

But the calculation is simpler if we observe that *a suitable rotation of the* Ox *and* Oy *axes about* O *converts equation* (1) *into an equation of the same form in which the complex numbers* a, b *are replaced by their moduli.*

In fact, let α, β be the arguments of a, b. Since

$$a = |\,a\,|\, e^{i\alpha}, \quad b = |\,b\,|\, e^{i\beta},$$

equation (1) can be written as

$$z = |\,a\,|\, e^{i(\alpha+\omega t)} + |\,b\,|\, e^{i(\beta-\omega t)}. \tag{2}$$

The vectors represented by the terms in the right member will have the same direction and the same sense for the particular value t_0 of t for which

$$\alpha + \omega t_0 = \beta - \omega t_0$$

or

$$t_0 = \frac{\beta - \alpha}{2\omega}.$$

Let us rotate the axes through the angle

$$\varphi = \alpha + \omega t_0 = \tfrac{1}{2}(\alpha + \beta) = \beta - \omega t_0. \tag{3}$$

If z_1 is the new affix of Z for the system (Ox_1, Oy_1), we have (24)

$$z = z_1 e^{i\varphi}$$

and equation (2) becomes

$$z_1 = |a| e^{i(\alpha+\omega t-\varphi)} + |b| e^{i(\beta-\omega t-\varphi)}$$

or, because of equations (3),

$$z_1 = |a| e^{i\omega(t-t_0)} + |b| e^{-i\omega(t-t_0)}.$$

By introducing the new real parameter

$$t_1 = t - t_0$$

we obtain the announced form

$$z_1 = |a| e^{i\omega t_1} + |b| e^{-i\omega t_1}. \tag{4}$$

In this equation, the separation of the real and imaginary parts gives

$$x_1 = [|a| + |b|] \cos \omega t_1,$$

$$y_1 = [|a| - |b|] \sin \omega t_1,$$

and the locus of Z, thus having

$$\frac{x_1^2}{[|a| + |b|]^2} + \frac{y_1^2}{[|a| - |b|]^2} = 1$$

for its cartesian equation, is an ellipse (E) of center O.

Remark. If $|a| = |b|$, equation (4) becomes

$$z_1 = 2|a| \cos \omega t_1.$$

This represents the segment of length $4|a|$ lying on Ox_1 and having O for midpoint, and is described by point Z_1 with a simple harmonic motion.

47. Construction of the elements of the ellipse. 1° *The tangent at Z is perpendicular to the line A_tB_t, which is parallel to the normal ZN at Z of the ellipse* (E).

The tangent ZT to (E) at Z is parallel to the vector representing the complex number (34)

$$\frac{dz}{dt} = i\omega (ae^{i\omega t} - be^{-i\omega t})$$

obtained by differentiating equation (1).

But the number

$$ae^{i\omega t} - be^{-i\omega t}$$

is represented by $\overline{OA_t} - \overline{OB_t}$ or $\overline{B_tA_t}$, whence the direction of the vector representing dz/dt is perpendicular to that of $\overline{B_tA_t}$.

2° *The major axis of* (E) *lies on the interior bisector* Ox_1 *of angle* $(\overline{OA_t}, \overline{OB_t})$; *the semi-major axis is* $|OA| + |OB|$, *and,* if $|OA| > |OB|$, *the semi-minor axis is* $|OA| - |OB|$.

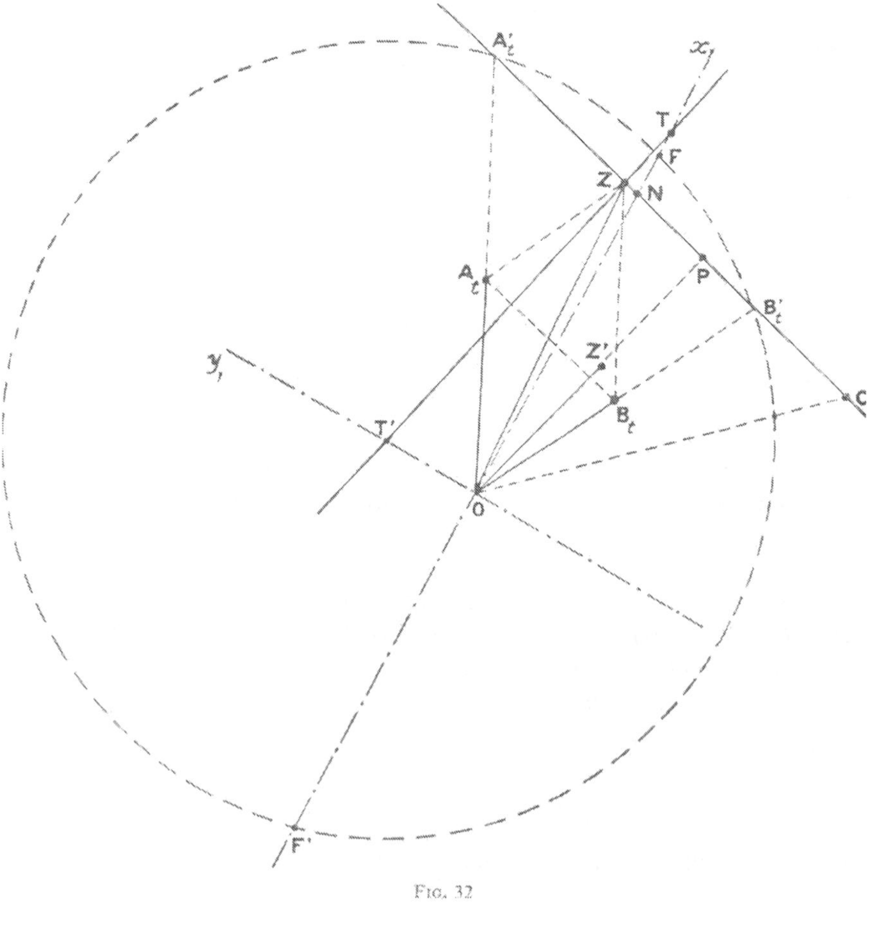

FIG. 32

As t varies, the lines bearing the vectors $\overline{OA_t}$, $\overline{OB_t}$ generate two pencils which, being inversely equal, have in common two perpendicular rays Ox_1, Oy_1, the interior and exterior bisectors of every angle $(\overline{OA_t}, \overline{OB_t})$. When $\overline{OA_t}$, $\overline{OB_t}$ are on Ox_1, they have the same sense, while their senses are opposite when they are on Oy_1. In each case, the tangent to (E) at the point Z which these vectors determine is perpendicular to A_tB_t (see 1°), that is, to the diameter OZ, and the latter is thus an axis of the ellipse.

3° *If the lines* OA_t, OB_t *meet the normal* ZN *to* (E) *in* A'_t, B'_t, *the foci of the ellipse are the intersections* F, F′ *of the* Ox_1 *axis with the circle passing through* A'_t, B'_t *and having its center at the point of intersection* T′ *of the* Oy_1 *axis and the tangent* ZT.

The affixes of these foci are the two square roots of the product 4ab.

We have, by 2°,

$$OF^2 = OF'^2 = [|OA| + |OB|]^2 - [|OA| - |OB|]^2 = 4\,|OA.OB|$$
$$= |OA'_t| \cdot |OB'_t|.$$

Since, moreover, Ox_1 is the interior bisector of angle $(\overline{OA_t}, \overline{OB_t})$, we have (**30, III**)

$$(A'_tB'_tFF') = -1$$

whence (**31,** 1) the indicated construction for F and F′ and, if f is the affix of F,

$$f^2 = a'_tb'_t = 2a_t \,.\, 2b_t = 4ab.$$

We also note that if T and N are the points of Ox_1 on the tangent and the normal at Z, then $(TNFF') = -1$, $OF = -OF'$, from which we get another construction of F and F′.

4° *The semi-diameter* OZ′ *conjugate to* OZ *is perpendicular to* A_tB_t *and has* $|A_tB_t|$ *for length.*

The diameter conjugate to OZ is parallel to the tangent ZT, and therefore perpendicular to A_tB_t. Since the tangent is parallel to the vector represented by the number (see 1°)

$$i(ae^{i\omega t} - be^{-i\omega t}) = e^{i\pi/2}.ae^{i\omega t} + e^{-i\pi/2}.be^{-i\omega t} = ae^{i\omega(t+\pi/2\omega)} + be^{-i\omega(t+\pi/2\omega)}$$

and since Z′ must be furnished by a value t' of t giving

$$z' = ae^{i\omega t'} + be^{-i\omega t'},$$

we conclude that Z′ is obtained for

$$t' = t + \frac{\pi}{2\omega}$$

and that

$$| OZ' | = | ae^{i\omega t'} + be^{-i\omega t'} | = | ae^{i\omega t} - be^{-i\omega t} | = | A_tB_t |.$$

5° *The center of curvature* C *of the ellipse for the point* Z *is the harmonic conjugate, with respect to the points* A'_t, B'_t, *of the orthogonal projection* P *of the center of the ellipse on the normal.*

At the instant t, the magnitude v of the velocity of the point Z describing the ellipse is (1°)

$$v = \left| \frac{dz}{dt} \right| = | \omega A_tB_t |$$

while the acceleration vector is represented by the complex number

$$\frac{d^2z}{dt^2} = -\omega^2 (ae^{i\omega t} + be^{-i\omega t}) = -\omega^2 z.$$

This vector is therefore $\omega^2\overline{ZO}$. Consequently, the algebraic value of the normal acceleration, on the arbitrarily oriented normal ZN, is $ZP.\omega^2$. But we know that $ZP.\omega^2$ has the value

$$\frac{v^2}{ZC} = \frac{\omega^2 A_tB_t^2}{ZC} = \frac{\omega^2 ZB_t'^2}{ZC}$$

whence we have

$$ZB_t'^2 = ZP.ZC,$$

which establishes the property.

The point C can be constructed by noticing that OP and OC cut off on ZB_t a segment of which B_t is the midpoint.

Corollary. *The radius vector* OZ *sweeps over an area which increases proportionally to the time* t.

We have just seen that under the motion of equation (1), the acceleration $\omega^2\overline{ZO}$ always passes through the fixed point O, and the planar motion takes place under the action of a central force.

48. Theorem. *Every ellipse can be generated with the aid of two determined rotating vectors.*

Consider (see Fig. 32) an ellipse (E) with foci F, F′, and whose minor axis lies on Oy_1. If the tangent to (E) at a point Z cuts Oy_1 in T′, let A'_t, B'_t be the points where the normal at Z is cut by the circle having center T′ and passing through F, F′. The vectors $\overline{OA'_t}/2$, $\overline{OB'_t}/2$ are the sought vectors $\overline{OA_t}$, $\overline{OB_t}$, which must rotate with angular velocities that are equal and opposite, but of arbitrary magnitude inasmuch as the lengths of the axes of the generated ellipse are independent of ω.

If we know, on Ox_1 and Oy_1, the vertices X and Y of (E), the sought vectors $\overline{OA_t}$, $\overline{OB_t}$ are carried by two arbitrary axes symmetric with respect to Ox_1 and, when $|OX| > |OY|$, the moduli of these vectors are (**47**, 2°)

$$\frac{|OX| + |OY|}{2}, \quad \frac{|OX| - |OY|}{2}.$$

49. Ellipse, hypocycloidal curve. Theorem I. *An ellipse is in two ways the locus of a point Z which describes a circle with a constant angular velocity, while the circle rotates around a fixed point O with a constant angular velocity equal to minus half the first angular velocity.*

It suffices to interpret equation (1) of the ellipse (E) written in one of the forms

$$z = ae^{i\omega t} + be^{i\omega t}.e^{-2i\omega t}, \tag{5}$$

$$z = ae^{-i\omega t}.e^{2i\omega t} + be^{-i\omega t}. \tag{6}$$

Consider equation (5). Let A, B be the points with affixes a, b; the fourth vertex Z_0 of the parallelogram constructed on OA and OB is the position of the moving point at the instant $t=0$. By rotation about O with constant angular velocity ω, the points A, Z_0, and the circle (c_0) of center A and radius $|AZ_0|$ occupy at instant t the positions A_t, Z′, (c_t). The affix of A_t is $ae^{i\omega t}$, and vector $\overline{A_tZ'}$ is obtained from $\overline{AZ_0}$ by the rotation represented by the number

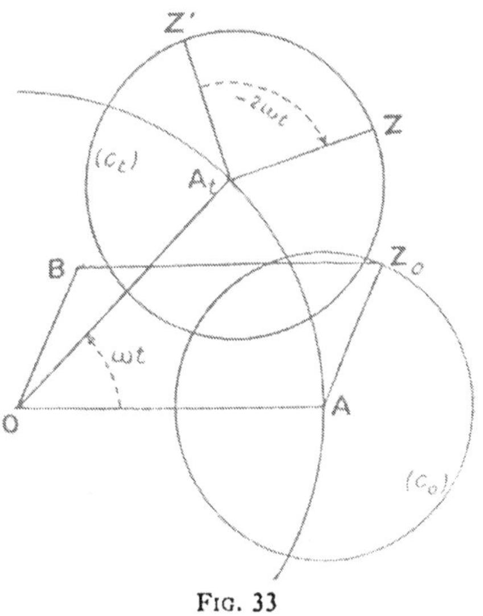

FIG. 33

$be^{i\omega t}$. If, then, we rotate $\overline{A_tZ'}$ about A_t through the angle $-2\omega t$, we have the vector $\overline{A_tZ}$ representing the number $be^{i\omega t}.e^{-2i\omega t}$, and the point Z can then be obtained as stated in the theorem.

Equation (6) gives Z by considering a point which describes the circle of center B and radius $|OA|$ with angular velocity 2ω, while this circle rotates about O with velocity $-\omega$.

Theorem II. *An ellipse is in two ways the hypocycloidal locus of a fixed point Z invariably connected with a circle which, without* *slipping, rolls interiorly on a circle of radius twice that of the first circle.*

At instant t, the motion of point Z on the ellipse (E) is tangent to the motion which results (*theorem I*) from the rotation of angular velocity -2ω about A_t and the rotation of angular velocity ω about O. These rotations compound into a single rotation about the instantaneous center of rotation C_a located on line OA_t and such that

$$\frac{OC_a}{C_aA_t} = \frac{-2\omega}{\omega} = -2.$$

The locus of C_a in the fixed plane, that is, in the plane of the base curve, is thus the circle β_a of center O and radius $|OC_a| = 2|OA_t| = 2|OA|$, while the locus of C_a in the moving plane, that is, in the plane of the generating curve, is the circle ρ_a of center A_t and radius $|A_tC_a| = |OA_t| = |OA|$. The ellipse (E) is then the locus of a point Z invariably connected with ρ_a as ρ_a rolls without slipping on β_a.

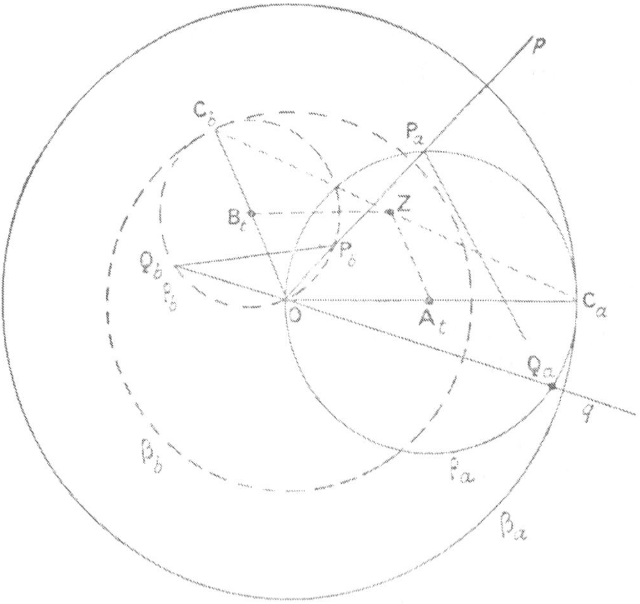

FIG. 34

If we reason in the same way about point B_t, we find that the instantaneous center of rotation C_b is such that

$$\frac{OC_b}{C_bB_t} = \frac{2\omega}{-\omega} = -2.$$

The base curve β_b and the generating curve ρ_b are the circles (O, 2 OB) and (B_t, OB).

Corollaries. 1° *Knowledge of* β_a, ρ_a, Z *implies knowledge of the rotating vectors* $\overline{OA}_t$, $\overline{OB}_t$, *with the aid of which* (E) *is generated* (**46**).

2° *The locus of a point* Z *on* ρ_a *and invariably connected with* ρ_a *is the diameter of* β_a *lying on* OZ (**46**, Remark).

Theorem III. *An ellipse is in two ways the locus of a point* Z *invariably connected with a segment of constant length whose extremities slide on two arbitrarily chosen diameters.*

In fact (Fig. 34), if two fixed straight lines p, q radiating from O cut the generating circle ρ_a in points P_a, Q_a, that we consider as invariably connected with ρ_a, point Z is invariably connected with the constant segment P_aQ_a, and we know (*theorem II, corollary* 2°) that the loci of P_a, Q_a are p, q when ρ_a rolls on β_a. We may reason in the same way with ρ_b, β_b.

V. *CYCLOIDAL CURVES*

50. The Bellermann-Morley generation with the aid of two rotating vectors. *If two vectors radiating from a fixed point* O *and having constant lengths rotate about* O *with constant and different angular velocities* ω_1, ω_2, *then the fourth vertex of the parallelogram having the two vectors for a pair of adjacent sides describes a cycloidal curve* (Γ).

Let Ox, Oy be two fixed rectangular axes, $\overline{OA_1}$, $\overline{OA_2}$ the initial positions of the rotating vectors, and a_1, a_2 the affixes of A_1, A_2.

At time t the vectors occupy positions $\overline{OA_{1t}}$, $\overline{OA_{2t}}$ such that the affixes of A_{1t}, A_{2t} are

$$a_1 e^{i\omega_1 t}, \quad a_2 e^{i\omega_2 t}.$$

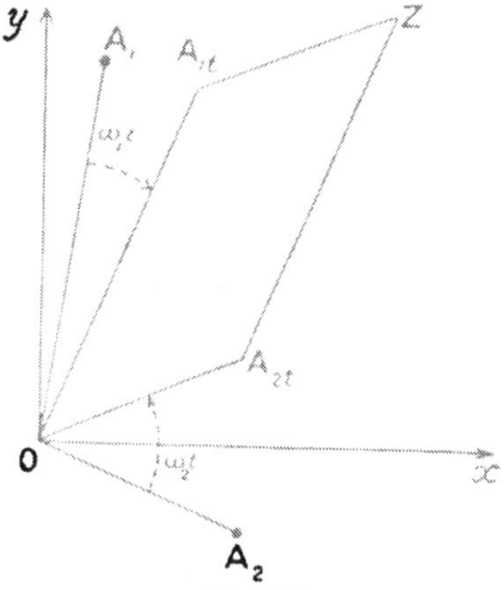

FIG. 35

The complex parametric equation of the cycloidal locus of the fourth vertex Z of the parallelogram constructed on OA_{1t}, OA_{2t} is then

$$z = a_1 e^{i\omega_1 t} + a_2 e^{i\omega_2 t}. \qquad (1)$$

We can show, as in article **46**, that *a suitable rotation of the* Ox *and* Oy *axes about* O *converts equation* (1) *into an equation of the same form in which the complex numbers* a_1, a_2 *are replaced by their moduli*, that is, into

$$z' = |a_1| e^{i\omega_1 t'} + |a_2| e^{i\omega_2 t'}.$$

In the new cartesian system (Ox', Oy'), the cartesian parametric equations of the cycloidal curve are then (**34**)

$$x' = |a_1| \cos \omega_1 t' + |a_2| \cos \omega_2 t',$$

$$y' = |a_1| \sin \omega_1 t' + |a_2| \sin \omega_2 t'.$$

51. Theorems. I. *A cycloidal curve is in two ways the locus of a point* Z *rotating uniformly about a point while that point rotates uniformly about a fixed point* O.

It suffices to interpret equation (1) written in one of the forms

$$z = a_1 e^{i\omega_1 t} + a_2 e^{i\omega_1 t} . e^{i(\omega_2 - \omega_1)t}, \tag{2}$$

$$z = a_1 e^{i\omega_2 t} . e^{i(\omega_1 - \omega_2)t} + a_2 e^{i\omega_2 t}, \tag{3}$$

by reasoning as in article **49**.

Equation (2) proves that the cycloidal curve is generated by a point which describes the circle of center A_1 and radius $|OA_2|$ with angular velocity $\omega_2 - \omega_1$, while this circle rotates about O with angular velocity ω_1.

Equation (3) says that the curve (Γ) is the locus of a point describing the circle of center A_2 and radius $|OA_1|$ with angular velocity $\omega_1 - \omega_2$, while at the same time this circle rotates about O with angular velocity ω_2.

II. *The cycloidal curve* (Γ) *is, in two ways, a roulette arising from a circular base curve and a circular generating curve.*

It is the path of a point invariably connected with

1° *a circle* ρ_1 *of center* A_1 *and radius* $|OA_1\omega_1/\omega_2|$ *which rolls without slipping on the circle* β_1 *of center* O *and radius* $|OA_1 (\omega_1 - \omega_2)/\omega_2|$;

2° *a circle* ρ_2 *of center* A_2 *and radius* $|OA_2\omega_2/\omega_1|$ *which rolls without slipping on the circle* β_2 *of center* O *and radius* $|OA_2 (\omega_1 - \omega_2)/\omega_1|$.

At an arbitrary instant, which can always be considered as the initial instant, the motion of point Z on (Γ) is tangent to the motion which results (*theorem I*) from the two rotations of centers O, A_1 and angular velocities ω_1, $\omega_2 - \omega_1$, or of centers O, A_2 and angular velocities ω_2, $\omega_1 - \omega_2$. These rotations compound into a single rotation about an instantaneous center C_1 or C_2 located on the line OA_1 or OA_2 and such that

$$\frac{OC_1}{C_1A_1} = \frac{\omega_2 - \omega_1}{\omega_1} \quad \text{or} \quad \frac{OC_2}{C_2A_2} = \frac{\omega_1 - \omega_2}{\omega_2}. \tag{4}$$

From this we obtain

$$\frac{OA_1}{C_1A_1} = \frac{\omega_2}{\omega_1}, \quad \frac{OA_2}{C_2A_2} = \frac{\omega_1}{\omega_2}. \tag{5}$$

In the moving plane, point C_1 is then at the constant distance $|A_1C_1| = |OA_1\omega_1/\omega_2|$ from the point A_1 invariably connected with this plane, and, in the fixed plane, at the constant distance $|OC_1| = |OA_1(\omega_1 - \omega_2)/\omega_2|$ from the fixed point O. Hence we have the generating circle ρ_1 and the base circle β_1.

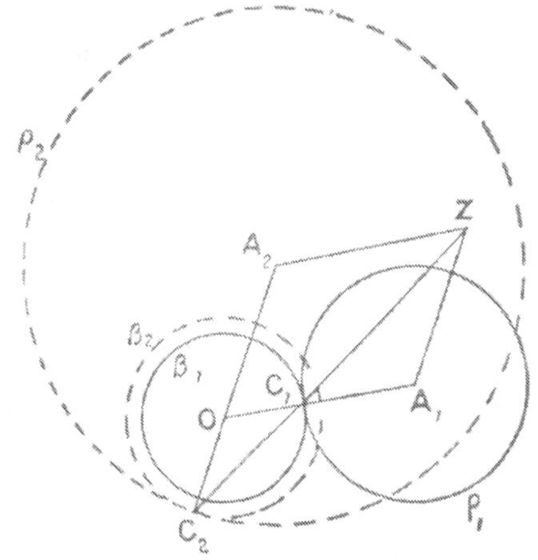

FIG. 36 $(\omega_1\omega_2 > 0)$

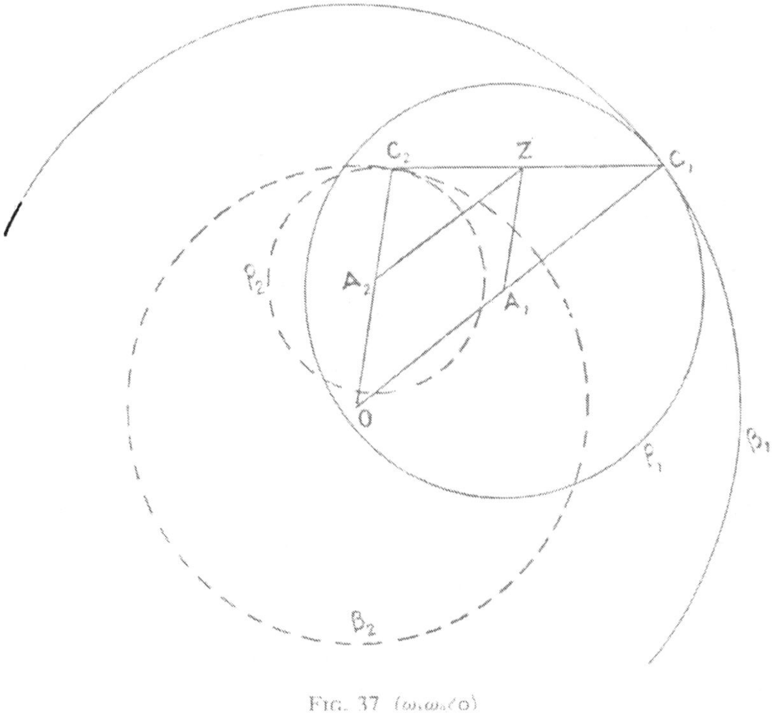

FIG. 37 $(\omega_1\omega_2 < 0)$

Corollaries. 1° *The normal at* Z *to* (Γ) *contains the instantaneous centers* C_1, C_2. Although this is known from kinematics, we shall reestablish it by a simple calculation. Since we have designated by A_1 the point of affix $a_1e^{i\omega_1 t}$, the affix of C_1 is

$$a_1e^{i\omega_1 t}\cdot\frac{OC_1}{OA_1}=a_1e^{i\omega_1 t}\cdot\frac{\omega_2-\omega_1}{\omega_2}$$

and the line ZC_1 has the direction of the vector represented by the number

$$(a_1e^{i\omega_1 t}+a_2e^{i\omega_2 t})-a_1e^{i\omega_1 t}\cdot\frac{\omega_2-\omega_1}{\omega_2}\quad\text{or}\quad\frac{1}{\omega_2}(a_1\omega_1e^{i\omega_1 t}+a_2\omega_2e^{i\omega_2 t}).$$

But the direction of the tangent at Z to (Γ) being, by (34),

$$\frac{dz}{dt}=i(a_1\omega_1e^{i\omega_1 t}+a_2\omega_2e^{i\omega_2 t}),$$

that of the normal is that of ZC_1 or, by similar reasoning, of ZC_2.

2° *If two points* C_1, C_2 *uniformly describe two concentric circles with angular velocities* ω_1, ω_2, *the line* C_1C_2 *remains normal to the cycloidal curve which is generated by the point* Z *of this line defined by the equation*

$$\frac{C_1Z}{C_2Z}=\frac{\omega_1}{\omega_2}.$$

This follows from

$$\frac{C_1Z}{C_2Z}=\frac{C_1A_1}{OA_1}$$

and one of the equations (5).

52. Epicycloids, hypocycloids. The cycloidal curve (Γ) is called an *epicycloid* or a *hypocycloid* according as the generating vectors $\overline{OA_1}$, $\overline{OA_2}$ rotate in the same or the opposite sense, that is, according as ω_1, ω_2 have or do not have the same sign.

For an epicycloid, the base circle and the generating circle are exteriorly or interiorly tangent, but in the latter case it is the larger circle which rolls on the smaller.

In fact, if $\omega_1\omega_2>0$ and, as in figure 36, $|\omega_1|<|\omega_2|$, equations (4) show that C_1 is between O and A_1, whence the circles ρ_1, β_1 touch exteriorly, while C_2 is on the prolongation of A_2O and the circles ρ_2, β_2 are tangent interiorly, the radius of the first being then greater than that of the second.

For a hypocycloid, the base circle and the generating circle are tangent interiorly, and it is the smaller circle which rolls on the larger.

In fact, if $\omega_1\omega_2 < 0$ and, as in figure 37, $|\omega_1| > |\omega_2|$, equations (5) show that C_1 is on the prolongation of OA_1, whence the circles ρ_1, β_1 are tangent interiorly and the first is the smaller. The same reasoning applies to ρ_2, β_2.

For a hypocycloid, the diameter of one generating circle is less than the radius of the corresponding base circle, while the diameter of the other generating circle is greater than the radius of the corresponding base circle.

In fact, since $\omega_1\omega_2$ is negative, equations (4) give

$$\left|\frac{OC_1}{A_1C_1}\right| = \left|\frac{\omega_2}{\omega_1}\right| + 1, \quad \left|\frac{OC_2}{A_2C_2}\right| = \left|\frac{\omega_1}{\omega_2}\right| + 1$$

and one of the right members of these is less than 2 while the other is greater than 2.

A cycloidal curve (epicycloid or hypocycloid) *is said to be ordinary if it is generated by a point of one of the generating circles. This point belongs to two generating circles, the two base circles coincide, the moduli of the rotating vectors are inversely proportional to those of their angular velocities, and the radii of the generating circles are proportional to the moduli of the angular velocities of the vectors which cause their centers to rotate.*

In fact, if Z is on the generating circle ρ_1 of radius **(51, II)** $|OA_1\omega_1/\omega_2|$, we have

$$|A_1Z| = |OA_2| = \left|OA_1\frac{\omega_1}{\omega_2}\right|,$$

then

$$|A_2Z| = |OA_1| = \left|OA_2\frac{\omega_2}{\omega_1}\right|$$

and Z is on the generating circle ρ_2. Moreover

$$\left|\frac{OA_1}{OA_2}\right| = \left|\frac{\omega_2}{\omega_1}\right|$$

whence the concentric base circles β_1, β_2 have **(51, II)** the same radius, and the ratio of the radii of the generating circles ρ_1, ρ_2 is

$$|A_1C_1| : |A_2C_2| = \left|OA_1\frac{\omega_1}{\omega_2}\right| : \left|OA_2\frac{\omega_2}{\omega_1}\right| = \left|\frac{\omega_1}{\omega_2}\right|.$$

An *epicycloid* is *lengthened* or *shortened* according as the point Z invariably connected with the generating circle exteriorly tangent to its base circle is exterior or interior to this generating circle.

A *hypocycloid* is *lengthened* or *shortened* according as the point Z invariably connected with the generating circle of diameter less than the radius of its base circle is exterior or interior to this generating circle.

In the two cases, *the cycloidal curve is lengthened or shortened if, supposing* $|\omega_1| < |\omega_2|$, *we have*

$$\left|\frac{OA_1}{OA_2}\right| < \left|\frac{\omega_2}{\omega_1}\right| \quad \text{or} \quad \left|\frac{OA_1}{OA_2}\right| > \left|\frac{\omega_2}{\omega_1}\right|.$$

The cycloidal curve of figure 36 ($\omega_1\omega_2 > 0$, $|\omega_1| < |\omega_2|$), which is an epicycloid, is lengthened if

$$|A_1Z| = |OA_2| > |A_1C_1|$$

or (51, II)

$$|OA_2| > \left|OA_1\frac{\omega_1}{\omega_2}\right| \quad \text{or} \quad \left|\frac{OA_1}{OA_2}\right| < \left|\frac{\omega_2}{\omega_1}\right|.$$

If for the hypocycloid we reason on figure 37, we should note that in this case $|\omega_1| > |\omega_2|$.[1]

VI. UNICURSAL CURVES

53. Definitions. A *plane curve* is said to be *unicursal* if it can be represented in cartesian coordinates by parametric equations

$$x = f_1(t), \quad y = f_2(t) \tag{1}$$

in which $f_1(t)$, $f_2(t)$ are *rational functions* of the real parameter t, that is, are polynomials or ratios of two polynomials in t.

Its complex parametric equation

$$z = f_1(t) + if_2(t)$$

then has the form

$$z = \frac{a_0 + a_1t + a_2t^2 + \ldots + a_mt^m}{b_0 + b_1t + b_2t^2 + \ldots + b_nt^n} = \frac{A(t)}{B(t)}, \tag{2}$$

the coefficients a, b of the polynomials A(t), B(t) of arbitrary degrees m, n being real or imaginary constants.

[1] In an article where these developments are found (*Ueber neue kinematische Modelle, sowie eine neue Einführung in die Theorie der cyclischen Curven*, Zeitschrift Math. Phys. XLIV, 1899, translated in the *Enseignement mathématique*, vol. 2, 1900, pp. 31-48), F. Schilling has given the construction of simple apparatuses based on these properties which permit all the cycloidal curves to be described mechanically.

Thus the straight line and the circle with respective equations (36, 41)

$$z = a_0 + a_1 t; \quad z = \frac{a_0 + a_1 t}{b_0 + b_1 t}, \quad a_0 b_1 - a_1 b_0 \neq 0, \quad \frac{b_0}{b_1} \text{ imaginary}$$

are unicursal curves.

54. Order of the curve. If we eliminate t from equations (1), by Sylvester's method, for example, we obtain the cartesian equation of the curve in the form

$$P(x, y) = 0,$$

where $P(x, y)$ is a polynomial of some degree s in the variables x and y.

The *curve* is thus *algebraic*, and we know that an arbitrary straight line cuts it in s points, real or imaginary, distinct or not. We say that the curve is of *order* s.

Theorem. *The unicursal curve with equation* (2) *in which*

1° *the polynomials* $A(t)$, $B(t)$ *are relatively prime, and*

2° $B(t)$ *vanishes for* ν *real or conjugate imaginary values (distinct or not), is of order*

$$m + n - \nu \quad \text{if} \quad m > n,$$
$$2n - \nu \quad \text{if} \quad m \leqslant n.$$

Let us first of all obtain from (2) the cartesian parametric equations. We must give to the fraction a denominator having real coefficients. Since the total number of real roots and of conjugate imaginary roots of the equation $B(t) = 0$ is ν, we can write

$$B(t) = B_\nu(t).B_{n-\nu}(t),$$

where $B_\nu(t)$ is a polynomial of degree ν whose coefficients are all real, while, if $\nu < n$, $B_{n-\nu}(t)$ is a polynomial of degree $n - \nu$ whose coefficients are not all real. The equation

$$B_{n-\nu}(t) = 0$$

has no real roots nor any pairs of conjugate imaginary roots. By grouping together the real and the pure imaginary parts we have

$$B_{n-\nu}(t) = B'(t) + iB''(t),$$

which we will write as

$$B_{n-\nu} = B' + iB'';$$

at least one of the polynomials B', B'' has degree $n - \nu$. Note that B', B'' are relatively prime, for if they should have a common root, real or imaginary, $B_{n-\nu}$ would have a real root or two conjugate imaginary roots. We similarly write

$$A(t) = A'(t) + iA''(t) \quad \text{or} \quad A = A' + iA'';$$

at least one of the polynomials A′, A″ has degree m. We then have

$$z = \frac{A' + iA''}{B_\nu(B' + iB'')}. \tag{3}$$

Multiplication of the numerator and denominator of the fraction by $B' - iB''$ gives the sought form

$$z = \frac{A'B' + A''B'' + i(A''B' - A'B'')}{B_\nu(B'^2 + B''^2)} \tag{4}$$

and consequently the cartesian parametric equations of the curve

$$x = \frac{A'B' + A''B''}{B_\nu(B'^2 + B''^2)}, \qquad y = \frac{A''B' - A'B''}{B_\nu(B'^2 + B''^2)}. \tag{5}$$

The real or imaginary points of the curve located on the arbitrary real or imaginary line with equation

$$y = px + q \tag{6}$$

correspond to the real or imaginary values of t which are roots of the equation

$$A''B' - A'B'' = p(A'B' + A''B'') + qB_\nu(B'^2 + B''^2)$$

obtained by replacing x and y in (6) by their values (5). This equation can be written as

$$qB_\nu(B'^2 + B''^2) + [A'(pB' + B'') + A''(pB'' - B')] = 0. \tag{7}$$

The degrees in t of the first and second terms are $2n - \nu$ and $m + n - \nu$, respectively.

The announced theorem will then be established if we show that no root of (7) gives the form 0/0 to one, and consequently to the other, of the expressions (5).

If a root t_0 of (7) renders x and y indeterminate, we will have

$$A'B' + A''B'' = 0,$$

$$A''B' - A'B'' = 0,$$

$$B_\nu(B' + iB'')(B' - iB'') = 0, \tag{8}$$

a system equivalent to equation (8) along with

$$A'B' + A''B'' + i(A''B' - A'B'') = 0 \text{ or } (A' + iA'')(B' - iB'') = 0, \tag{9}$$

$$A'B' + A''B'' - i(A''B' - A'B'') = 0 \text{ or } (A' - iA'')(B' + iB'') = 0. \tag{10}$$

By virtue of (9), t_0 annuls $A' + iA''$ or $B' - iB''$. Since the polynomials A, B are relatively prime, if t_0 annuls $A' + iA''$, this value also annuls, because of (10) and (8), $A' - iA''$ and $B' - iB''$; thus t_0 annuls each of the polynomials A', A'' and is imaginary inasmuch as B' and B'' are relatively prime; the number conjugate to t_0 therefore annuls A', A'', $B' + iB''$, and consequently the polynomials A, $B_{n-\nu}$, which is impossible. If, on the other hand, the value t_0 annuls $B' - iB''$, it is imaginary and does not annul $B' + iB''$, for B' and B'' are relatively prime. Hence, by virtue of (10), t_0 must annul $A' - iA''$; the conjugate of t_0 will then annul $B' + iB''$ and $A' + iA''$, which is impossible.

Remarks. 1° An *imaginary root* of (7) gives, with the aid of (5), an imaginary point of the curve and, by (4), a z whose image is not this point since each z has a real point for image (**1**).

2° *It may be that a unicursal curve consists of a curve of lower order but counted several times.*

Thus, according to the theorem, the equation

$$z = a_0 + a_1 t^2$$

represents a curve of the second order, or a conic. But if we set

$$t^2 = T, \quad T \geqslant 0$$

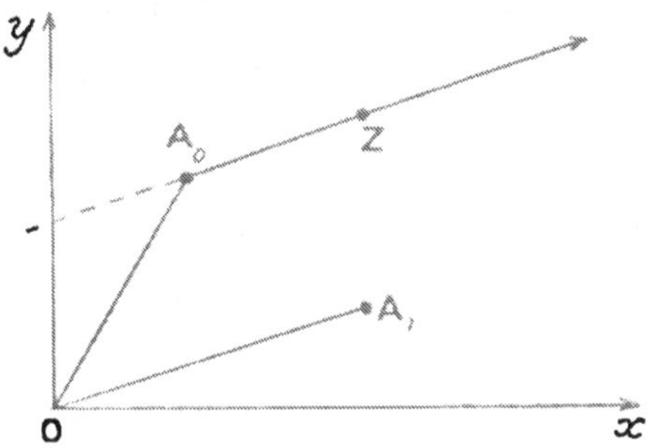

FIG. 38

the equation becomes

$$z = a_0 + a_1 T \,.$$

and represents (**36**) a ray radiating from the point A_0 of affix a_0, each point Z of which is given by two values of t, namely the two square roots of the number T which yields this point. As t varies from $-\infty$ to $+\infty$, the ray is thus described twice, namely in the two opposite senses, by the point Z.

Also notice the *case where* $B(t)$ *is the conjugate of* $A(t)$. Since $|z| = 1$, Z lies on a circle.

3° *The only non-irreducible fractions to which the theorem can be applied are obtained by multiplying the polynomials* $A(t)$, $B(t)$ *by the conjugates of an arbitrary number of linear factors into which the polynomial* $B_{n-\nu}(t)$ *resolves itself.*

In fact, if we multiply A(t) and B(t) by a common factor $t - \alpha$, the analogues of m, n, ν are

$$m_1 = m + 1, \quad n_1 = n + 1, \quad \nu_1$$

and we will have

$$m_1 + n_1 - \nu_1 = m + n - \nu, \quad 2n_1 - \nu_1 = 2n - \nu$$

if

$$\nu_1 = \nu + 2.$$

The only way of increasing by 2 the number of real or conjugate imaginary roots of $B(t) = 0$ by multiplying by the single factor $t - \alpha$ is to choose for α the conjugate of some one of the roots of $B_{n-\nu}(t) = 0$.

It follows that *the fraction (4) is the only fraction whose denominator polynomial has all its coefficients real and to which the theorem for the determination of the order of the curve can be applied.*

55. Point construction of the curve.[1] If we carry out the indicated division, equation (2) becomes

$$z = Q(t) + \frac{R(t)}{B(t)},$$

Q(t) being a polynomial

$$c_0 + c_1 t + c_2 t^2 + \ldots + c_{m-n} t^{m-n} \tag{11}$$

of degree $m - n$, and the fraction can be replaced by a sum of simple fractions in which a p-fold root α, real or imaginary, of $B(t) = 0$ gives rise to the partial sum

$$\sum_{k=0}^{k=p-1} \frac{l_k}{(t-\alpha)^{p-k}} \tag{12}$$

where the l_k's are constants,

The point Z corresponding to a value of t can then be found by obtaining the sum of the vectors representing the terms of (11) and those of the sums analogous to (12). For these latter it will be noted that

$$z = \frac{l_k^{\frac{1}{p-k}}}{t - \alpha}$$

is the equation of a straight line or of a circle passing through the origin ($t = \infty$) according as the number α is real or imaginary.

[1] G. Hauffe, *Ortskurven der Starkstromtechnik*, p. 131, Berlin, 1932.

56. Circular unicursal curves. A *plane algebraic curve* represented by a cartesian equation with real coefficients and which passes k times through one of the circular points at infinity also passes k times through the other circular point at infinity. The curve is said to be *k-circular;* it is called *circular, bicircular,* or *tricircular* according as k is 1, 2, or 3, that is, according as each of the circular points is a simple, double, or triple point of the curve.

A curve can be k-circular without being unicursal, and conversely, but it can also possess both of these characteristics.

Theorem. *The unicursal curve with equation* (54)

$$z = \frac{A(t)}{B(t)} = \frac{A(t)}{B_\nu(t) \, . \, B_{n-\nu}(t)} = \frac{A}{B_\nu(B' + iB'')} \qquad (13)$$

is $(n - \nu)$*-circular.*

The coordinates of an arbitrary point of the curve are given by (5). Since the circular points are at infinity, the values (imaginary) of t able to give a point of the curve which is to be a circular point must annul the denominator of the fractions (5) and thus figure among the roots of

$$B_\nu(B' + iB'')(B' - iB'') = 0. \qquad (14)$$

Moreover, the t's corresponding to the circular point situated on the isotropic line of equation

$$x + iy = 0 \qquad (15)$$

must be roots of

$$A'B' + A''B'' + i(A''B' - A'B'') = 0,$$

an equation which can be written as

$$(A' + iA'')(B' - iB'') = 0 \quad \text{or} \quad A(B' - iB'') = 0. \qquad (16)$$

The t's furnishing the considered circular point are therefore the roots common to equations (14), (16), or, since A and $B_\nu(B' + iB'')$ are two relatively prime polynomials, are the roots of

$$B' - iB'' = 0. \qquad (17)$$

This equation being of degree $n - \nu$, the theorem is established.

57. Foci. A real or imaginary point is a *focus* of a given plane curve if the two isotropic lines radiating from this point are tangents to the curve.

When the curve is k-circular, a *focus* is *singular* if the isotropic lines radiating from this point have their points of contact at the circular points. For example, the center of a circle is the only focus of this circular curve, and it is singular.

Theorem. *If the unicursal curve with equation* (56)

$$z = \frac{A}{B_\nu(B' + iB'')}$$

does not have a multiple point in the finite part of the plane on the isotropic lines radiating from the origin, then the origin is

1° *an ordinary focus if the equation* A $= 0$ *has a root which is at least double and which is not a root of*

$$B' - iB'' = 0; \tag{17}$$

2° *a singular focus if the equations in* t

$$A = 0, \quad B' - iB'' = 0$$

have at least one common root.

The t's of the points common to the curve and to the isotropic line of equation (15) are the roots of (16); the $n - \nu$ common circular points (coincident at the same circular point) correspond to the roots of (17) (56). Among the remaining common points, suppose that there are no two which coincide because of the existence on the isotropic line of a multiple point in the finite part of the plane. Then, in order that the origin be an ordinary focus, that is, in order that the isotropic line be tangent to the curve at a non-circular point, it is necessary and sufficient that two roots of the equation $A = 0$ be equal but be distinct from any root of $B' - iB'' = 0$. If, on the contrary, the origin is to be a singular focus, the point of tangency must lie at the circular point, and at least one of the points given by $A = 0$ must coincide with the circular point.

Determination of foci. A translation of axes to a sought focus of affix ϕ gives the curve the new equation

$$\zeta = \frac{A}{B_\nu(B' + iB'')} - \phi = \frac{A - \phi B_\nu(B' + iB'')}{B_\nu(B' + iB'')} = \frac{A_1}{B_\nu(B' + iB'')}.$$

The ϕ of an ordinary focus annuls the discriminant of A_1 and the double root then existing for $A_1 = 0$ must not annul $B' - iB''$. The ϕ of a singular focus annuls A_1 and $B' - iB''$. For the ϕ of an ordinary focus, it is necessary to see if the isotropic line contains or does not contain a multiple point in the finite part of the plane.

VII. CONICS

58. General equation. *The complex parametric equation of any real conic of the Gauss plane is of the form*

$$z = \frac{a_0 + 2a_1t + a_2t^2}{r_0 + 2r_1t + r_2t^2}, \tag{1}$$

where a_0, a_1, a_2 *are real or imaginary constants and* r_0, r_1, r_2 *are real constants.*

Consider a real conic referred to two perpendicular axes Ox, Oy and let ω be the affix of any one of its points Ω. For the new axes $\Omega\xi$, $\Omega\eta$ parallel, respectively, to Ox, Oy, the conic has an equation of the form

$$r_0\xi^2 + 2r_1\xi\eta + r_2\eta^2 - \alpha\xi - \beta\eta = 0, \tag{2}$$

r_0, r_1, r_2, α, β being real constants.

A variable real line through Ω and having equation

$$\eta = t\xi, \quad t \text{ a real parameter},$$

cuts the conic again in the point having cartesian coordinates

$$\xi = \frac{\alpha + \beta t}{r_0 + 2r_1t + r_2t^2}, \quad \eta = \frac{\alpha t + \beta t^2}{r_0 + 2r_1t + r_2t^2}$$

and whose affix, for the system $\Omega\xi$, $\Omega\eta$, is

$$\zeta = \xi + i\eta = \frac{\alpha + (\beta + i\alpha)t + i\beta t^2}{r_0 + 2r_1t + r_2t^2}. \tag{3}$$

The affix z of this point, for the system Ox, Oy, is then **(24)**

$$z = \zeta + \omega = \frac{\alpha + \omega r_0 + (\beta + i\alpha + 2\omega r_1)t + (i\beta + \omega r_2)t^2}{r_0 + 2r_1t + r_2t^2}, \tag{4}$$

an equation of the form (1).

Particular cases.

$$z = a_0 + 2a_1t + a_2t^2, \tag{5}$$

$$z = \frac{a_0 + 2a_1t + a_2t^2}{r_0 + 2r_1t}. \tag{6}$$

59. Species. *Equation* (1) *represents an ellipse, a hyperbola, or a parabola according as the discriminant*

$$\Delta_r = r_0 r_2 - r_1^2$$

of the trinomial

$$r_0 + 2r_1 t + r_2 t^2 \tag{7}$$

is positive, negative, or zero.

For the conic is an ellipse, a hyperbola, or a parabola according as there exist 0, 2, or 1 real values of t which make z infinite.

The only value of t which makes z of equation (5) infinite is $t = \infty$; we then have a parabola and, since $r_1 = r_2 = 0$, we have $\Delta_r = 0$.

The z given by (6) is infinite if t is infinite or if $t = -r_0/2r_1$; we then have a hyperbola and, since $r_2 = 0$, we have $\Delta_r = -r_1^2 < 0$.

Corollary. *Equation* (1) *represents a circle if,*

$$\Delta_a = a_0 a_2 - a_1^2, \quad \mathrm{H} = a_0 r_2 - 2a_1 r_1 + a_2 r_0$$

being respectively the discriminant of the trinomial

$$a_0 + 2a_1 t + a_2 t^2 \tag{8}$$

and the harmonic invariant of the trinomials (7), (8), *we have*

$$\Delta_r > 0, \quad 4\,\Delta_a \Delta_r - \mathrm{H}^2 = 0. \tag{9}$$

In fact, relations (9) say that the zeros of (7) are imaginary and that one of them is a zero of (8), whence (1) can be written in the form

$$z = \frac{a_0 + a_1 t}{b_0 + b_1 t}, \quad \frac{b_1}{b_0} \text{ imaginary},$$

and represents a circle (**41**).

60. Foci, center. 1° *The affixes* ϕ_1, ϕ_2 *of the foci of the conic with equation* (1) *are the roots of the equation*

$$\Delta_r \phi^2 - \mathrm{H}\phi + \Delta_a = 0. \tag{10}$$

In fact, if we translate the axes to a focus of affix ϕ, the equation of the conic becomes

$$z_1 = \frac{a_0 - \phi r_0 + 2(a_1 - \phi r_1)\,t + (a_2 - \phi r_2)\,t^2}{r_0 + 2r_1 t + r_2 t^2}.$$

Since the ellipse, the hyperbola, and the parabola are not circular curves, their foci are ordinary, and the numerator of the fraction must be the square of a linear function of t (**57**), whence

$$(a_0 - \phi r_0)(a_2 - \phi r_2) - (a_1 - \phi r_1)^2 = 0,$$

which is equation (10).

When the conic is a parabola, the affix of the focus is

$$\phi_1 = \frac{\Delta_a}{\mathrm{H}}.$$

2° *The affix ω of the center of the conic is*

$$\omega = \frac{\mathrm{H}}{2\Delta_r}$$

since the center is the midpoint of the segment determined by the foci, and we have $\omega = (\phi_1 + \phi_2)/2$.

Corollary. *The origin is a focus or the center of the conic according as*

$$\Delta_a = 0 \quad \text{or} \quad \mathrm{H} = 0.$$

61. Center and radius of a circle. *The affix of the center of the circle of equation* (**41**)

$$z = \frac{at + b}{ct + d} \tag{11}$$

is

$$\omega = \frac{a\bar{d} - b\bar{c}}{c\bar{d} - d\bar{c}} \tag{12}$$

and the radius of the circle is

$$\left| \frac{ad - bc}{c\bar{d} - d\bar{c}} \right|.$$

Since the center is a singular focus of the circle, the numerator of (**57**)

$$\zeta = \frac{at + b - \omega(ct + d)}{ct + d}$$

must vanish for the root of

$$\bar{c}t + \bar{d} = 0,$$

and therefore

$$-a\frac{\bar{d}}{\bar{c}} + b - \omega\left(-c\frac{\bar{d}}{\bar{c}} + d\right) = 0,$$

whence the expression (12).[1]

[1] H. Pflieger-Haertel (*Zur Theorie der Kreisdiagramme: Archiv für Elektrotechnik*, vol. XII, 1923, pp. 486-493) has calculated ω by using some properties of a Möbius transformation (see article **127**).

The affix z of any point of the circle gives the radius R by

$$R = |\omega - z|.$$

If we take $z = b/d$, corresponding to $t = 0$, we have

$$R = \left| \frac{a\bar{d} - b\bar{c}}{c\bar{d} - d\bar{c}} - \frac{b}{d} \right| = \left| \frac{ad\bar{d} - bc\bar{d}}{d(c\bar{d} - d\bar{c})} \right| = \left| \frac{ad - bc}{c\bar{d} - d\bar{c}} \right|.$$

62. Parabola. 1° Consider the equation (58)

$$z = a_0 + 2\,a_1 t + a_2 t^2. \tag{13}$$

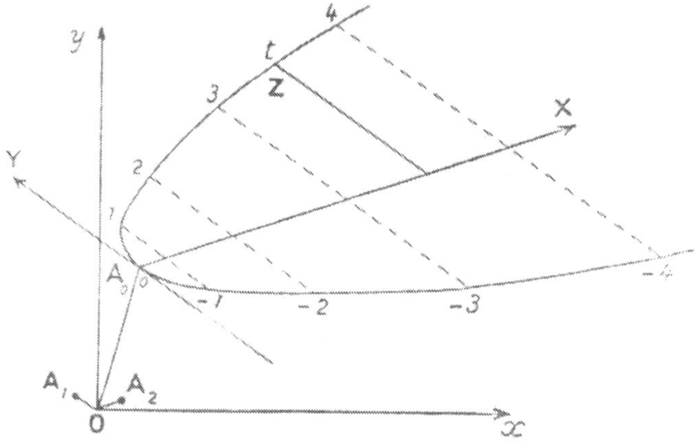

FIG. 39

Let A_0, A_1, A_2 be the points with affixes a_0, a_1, a_2 and let A_0X, A_0Y be axes having the directions of the vectors $\overline{OA}_2$, $\overline{OA}_1$. In the cartesian system A_0X, A_0Y, the coordinates of a point Z of the curve are

$$X = |a_2|\,t^2, \quad Y = 2\,|a_1|\,t$$

and the equation of the parabola is

$$Y^2 = 4\,\frac{|a_1|^2}{|a_2|}\,X.$$

We see that *the numbers* a_0, a_2, a_1 *give a point* A_0 *of the curve, the direction of the diameters of the curve, and the direction of the tangent to the curve at* A_0. As for the focus, its affix is (60)

$$a_0 - \frac{a_1^2}{a_2}.$$

Therefore (30, III), *if* A_3 *is the harmonic conjugate of* A_2 *with respect to point* A_1 *and the symmetric of* A_1 *with respect to* O, *then the focus* F *of the parabola is defined by* $\overline{A_0F} = \overline{A_3O}$. We can easily find the other fundamental elements of the curve.

2° Consider the equation (59)

$$z = \frac{a_0 + 2a_1t + a_2t^2}{r_0 + 2r_1t + r_2t^2}, \quad \Delta_r = r_0r_2 - r_1^2 = 0. \qquad (14)$$

It assumes the form (13) by the substitution

$$t = \frac{r_1}{r_2}\frac{T}{1-T}$$

and becomes

$$z = \frac{1}{r_1^2}[a_0r_2 - 2(a_0r_2 - a_1r_1)T + (a_0r_2 - 2a_1r_1 + a_2r_0)T^2].$$

It follows that *the diameters are parallel to the vector which represents the harmonic invariant* $H = a_0r_2 - 2a_1r_1 + a_2r_0$, and *the parametric equation of the axis of the curve is* (60)

$$z = \frac{\Delta_a}{H} + Ht_1.$$

3° *When the origin is the focus of the parabola,* we have $\Delta_a = 0$ and equation (14) becomes

$$z = \left(\sqrt{\frac{r_2}{a_2}}\,\frac{a_1 + a_2t}{r_1 + r_2t}\right)^2.$$

63. Hyperbola. The equation of any hyperbola of the plane is (58, 59)

$$z = \frac{a_0 + 2a_1t + a_2t^2}{r_0 + 2r_1t + r_2t^2}, \quad \Delta_r = r_0r_2 - r_1^2 < 0 \qquad (15)$$

and the affix of its center is (60)

$$\omega = \frac{H}{2\Delta_r} = \frac{a_0r_2 - 2a_1r_1 + a_2r_0}{2\Delta_r}. \qquad (16)$$

Case 1: $r_0 = r_2 = 0$. We have $\omega = a_1/r_1$ and the equation becomes

$$z = \omega + \frac{a_2}{2r_1}t + \frac{a_0}{2r_1t}. \qquad (17)$$

The infinite points of the curve correspond to the values ∞ and 0 of t. Since the vector $\overline{OZ}$ has the same direction as $(1/t)\,\overline{OZ}$ and $t\overline{OZ}$, the corresponding asymptotes d, d' are parallel to the vectors $\overline{OD}$, $\overline{OD}'$ representing the numbers $a_2/2r_1$, $a_0/2r_1$, or, since

r_1 is real, parallel to the vectors which represent a_2, a_0. Consequently, if we mark off on the asymptotes from the center Ω :

$$\overline{\Omega D_1} = t\ \overline{OD}, \quad \overline{\Omega D_1'} = \frac{1}{t}\ \overline{OD'},$$

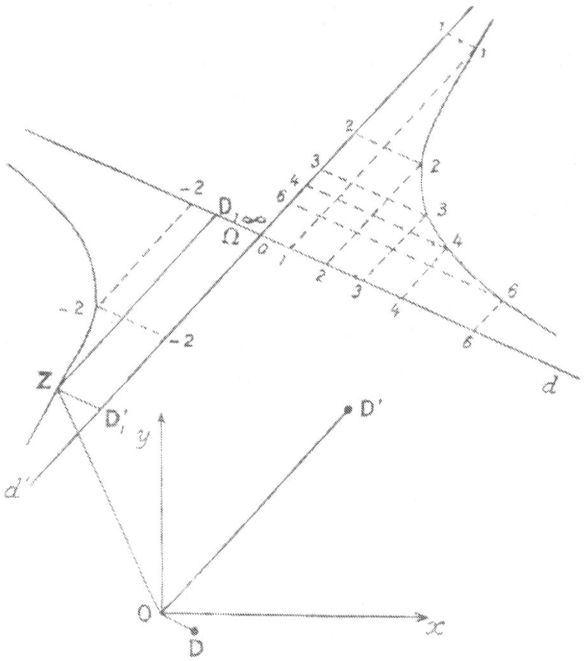

FIG. 40

the fourth vertex of the parallelogram constructed on ΩD_1, $\Omega D_1'$ is a point Z of the hyperbola. The hyperbola is thus easily calibrated in this way.

Case 2: r_0 *and* r_2 *not both zero.* This case can be reduced to the preceding by substituting for t a real parameter T related to t by

$$t = \frac{\alpha T + \beta}{\gamma T + \delta}, \quad \alpha\delta - \beta\gamma \neq 0, \quad \alpha, \beta, \gamma, \delta \text{ real.} \tag{18}$$

Equation (15) becomes

$$z = \frac{a_0\delta^2 + 2\,a_1\beta\delta + a_2\beta^2 + 2[a_0\gamma\delta + a_1(\alpha\delta + \beta\gamma) + a_2\alpha\beta]\,T + (a_0\gamma^2 + 2\,a_1\alpha\gamma + a_2\alpha^2)\,T^2}{r_0\delta^2 + 2\,r_1\beta\delta + r_2\beta^2 + 2[r_0\gamma\delta + r_1(\alpha\delta + \beta\gamma) + r_2\alpha\beta]\,T + (r_0\gamma^2 + 2\,r_1\alpha\gamma + r_2\alpha^2)\,T^2} \tag{19}$$

and will have the form of the first case if

$$r_0\gamma^2 + 2r_1\gamma\alpha + r_2\alpha^2 = 0, \tag{20}$$

$$r_0\delta^2 + 2r_1\delta\beta + r_2\beta^2 = 0. \tag{21}$$

Suppose $r_0 \neq 0$. Since $\Delta_r < 0$, the equation

$$r_0\xi^2 + 2r_1\xi + r_2 = 0 \tag{22}$$

has two real distinct roots ξ_1, ξ_2 and, by virtue of the inequality (18), the relations (20), (21) give, for example,

$$\frac{\gamma}{\alpha} = \xi_1, \quad \frac{\delta}{\beta} = \xi_2.$$

If we should have $r_0 = 0$, $r_2 \neq 0$, we would consider the equation

$$r_2\zeta^2 + 2r_1\zeta + r_0 = 0$$

instead of (22).

For the substitution

$$t = \frac{T+1}{\xi_1 T + \xi_2}$$

corresponding to

$$\alpha = 1,\ \beta = 1,\ \gamma = \xi_1,\ \delta = \xi_2,$$

equation (19) becomes

$$z = \omega + \frac{r_0}{4\Delta_r}(a_0\xi_1^2 + 2a_1\xi_1 + a_2)\,T + \frac{r_0}{4\Delta_r}(a_0\xi_2^2 + 2a_1\xi_2 + a_2)\cdot\frac{1}{T}.$$

64. Ellipse. *The ellipse with equation*

$$z = \frac{a_0 + 2a_1t + a_2t^2}{r_0 + 2r_1t + r_2t^2}, \quad \Delta_r > 0 \tag{23}$$

can be constructed with the aid of two calibrated circles passing through the origin.

By carrying out the division indicated by the fraction, we have

$$z = \frac{a_2}{r_2} + \frac{2(a_1r_2 - a_2r_1)\,t + a_0r_2 - a_2r_0}{(r_0 + 2r_1t + r_2t^2)\,r_2}.$$

If $\alpha + i\beta$, $\alpha - i\beta$ are the zeros of the denominator, the fraction in t decomposes into

$$\frac{A}{t - \alpha - i\beta} + \frac{B}{t - \alpha + i\beta}$$

where

$$A = \frac{2(a_1r_2 - a_2r_1)(\alpha + i\beta) + a_0r_2 - a_2r_0}{2i\beta r_2^3},$$

and where B is obtained from A by replacing i by $-i$. Since the equations

$$z = \frac{A}{t - \alpha - i\beta}, \quad z = \frac{B}{t - \alpha + i\beta}$$

are those of two circles passing through the origin, the statement is established.

Remark. The equation

$$z = ae^{i\omega t} + be^{-i\omega t} \tag{24}$$

of an ellipse generated by two rotating vectors (**46**) reduces to the form (23) if we set

$$\tan\frac{\omega t}{2} = T$$

inasmuch as

$$\cos\omega t = \frac{1 - T^2}{1 + T^2}, \quad \sin\omega t = \frac{2T}{1 + T^2},$$

$$z = (a + b)\cos\omega t + i\,(a - b)\sin\omega t = \frac{a + b + 2i(a - b)\,T - (a + b)\,T^2}{1 + T^2}.$$

Conversely, by utilizing (18) we can obtain an equation of the form (24) from equation (23). We leave to the reader the task of carrying this through.

Exercises 46 through 59

46. An ellipse being given by the positions of two conjugate diameters, construct the two rotating vectors which generate the ellipse and construct the axes of the curve.

47. 1° Two points Z_1, Z_2 describe two circles of center O with angular velocities ω, $-\omega$. Study the locus of the point which divides segment Z_1Z_2 in a given ratio and examine its degenerate cases.

2° The locus of the harmonic conjugate of O with respect to Z_1, Z_2 is a unicursal bicircular quartic which is the transform by inversion of an ellipse whose center is at the center of inversion.

3° Consider 1° when the circles are not concentric.

48. The parabola with equation (13) of article **62** degenerates into two coincident half-lines if $a_1 = 0$. When $a_1 \neq 0$:

1° determine the tangent at A_0 by article **34**, and the direction of the diameters by dividing by t^2;

2° the parametric equation of the axis of the parabola is

$$z = a_0 - \frac{a_1^2}{a_2} + a_2 t_1;$$

3° the vertex S of the parabola is given by the value

$$\frac{-(a_1\bar{a}_2 + \bar{a}_1 a_2)}{2a_2\bar{a}_2}$$

of the parameter t, and has for affix

$$s = a_0 - \frac{a_1^2}{a_2} + \frac{(a_1\bar{a}_2 - \bar{a}_1 a_2)^2}{4a_2\bar{a}_2^2}$$

[apply article 10];

4° calculate the affix of the foot of the directrix and show that its distance from the focus is

$$\frac{|a_2| \cdot \left|\frac{a_1}{a_2} - \frac{\bar{a}_1}{\bar{a}_2}\right|^2}{2} \quad \text{or} \quad 8\,\frac{OA_1^2}{OA_2}\sin^2(\overline{OA_1},\overline{OA_2}).$$

49. The parabola having equation (14) of article **62** degenerates into two coincident half-lines if $a_0/r_0 = a_1/r_1$. [See exercise 48.]

50. Let us be given two fixed points A, B and a line d which does not pass through A. If D is a point moving on d, let Z be the point on the line symmetric to AB with respect to AD such that

$$AZ \cdot AB = AD^2.$$

Find the locus of Z. [Take the origin at A and use articles **8** and **62** and exercise 48.]

51. If t is a real parameter, the equation

$$z = at + \frac{b}{t}$$

represents :

1° a straight line counted twice, but from which we omit a finite segment centered at O, if a and b have the same argument;

2° a straight line counted twice if the arguments of a and b differ by π;

3° a hyperbola in all other cases, and which is equilateral if

$$a\bar{b} + \bar{a}b = 0.$$

The foci have $\pm 2\sqrt{ab}$ for affixes; construct them [article 31]. By observing that at a vertex the tangent is normal to the diameter ending at the point of contact, show that the real vertices are given by

$$t = \pm\sqrt[4]{\frac{b\bar{b}}{a\bar{a}}}.$$

52. The hyperbola with equation (15) of article **63** is equilateral if (exercise 51)

$$a_0[\bar{a}_0 r_2^2 - 2\bar{a}_1 r_2 r_1 + \bar{a}_2(2r_1^2 - r_0 r_2)] - 2a_1[\bar{a}_0 r_1 r_2 - 2\bar{a}_1 r_0 r_2 + \bar{a}_2 r_0 r_1]$$
$$+ a_2[\bar{a}_0(2r_1^2 - r_0 r_2) - 2\bar{a}_1 r_0 r_1 + \bar{a}_2 r_0^2] = 0.$$

The hyperbola degenerates into two coincident lines if

$$\begin{vmatrix} a_0 & a_1 & a_2 \\ \bar{a}_0 & \bar{a}_1 & \bar{a}_2 \\ r_0 & r_1 & r_2 \end{vmatrix} = 0.$$

[Show that the quotient of the coefficients of T and 1/T (article 63) is real.] A segment is or is not omitted according as

$$a_0\bar{a}_0 r_2^2 + 4a_1\bar{a}_1 r_2 r_0 + a_2\bar{a}_2 r_0^2 - 2(a_0\bar{a}_1 + \bar{a}_0 a_1) r_1 r_2$$
$$- 2(a_1\bar{a}_2 + \bar{a}_1 a_2) r_0 r_1 + (a_0\bar{a}_2 + \bar{a}_0 a_2)(2r_1^2 - r_0 r_2)$$

is positive or negative. [Add to the quotient of the coefficients of T and 1/T the conjugate of the quotient.]

53. Squares of the moduli of the roots of a second degree equation. 1° Show that the moduli ζ_1, ζ_2 of the roots of

$$z^2 - 2pz + q = 0,$$

where p and q are real or imaginary, have for squares the roots of

$$x^2 - 2[|p^2| + |p^2 - q|]x + |q^2| = 0.$$

Write this equation when p and q are real.

2° Deduce that $\zeta_1 = \zeta_2$ if $p = 0$ or if q/p^2 is a real number not less than 1. [If

$$a^2 = p^2 - q,$$

we have

$$\zeta_1 = |p + a|,\quad \zeta_1^2 = (p + a)(\bar{p} + \bar{a}),\quad \zeta_2^2 = ...,\quad \zeta_1^2 + \zeta_2^2 = ...,\quad \zeta_1\zeta_2 = |q|,\quad ....$$

If p and q are real, treat the cases where $p^2 - q$ is positive, zero, or negative. For 2°, observe that the equation

$$|m| + |n| = |m + n|$$

can hold only if $mn = 0$ or if m/n is a positive real number.]

54. The curve with equation

$$z = \frac{4t^2 + 2(2 + i)t + i}{2t^2 + 2t + 1}$$

is an ellipse whose foci and vertices have the affixes $\pm\sqrt{3}$, ± 2, $\pm i$.

55. 1° If z_0 is the affix of an arbitrary point of the ellipse with equation

$$z = \frac{a_0 + 2a_1 t + a_2 t^2}{r_0 + 2r_1 t + r_2 t^2}$$

and with center assumed at the origin (H = 0, article 60), the ellipse can also be represented with a parameter t_1 by the equation

$$z = ae^{it_1} + be^{-it_1}, \tag{1}$$

in which a, b are the roots of

$$u^2 - z_0 u - \frac{\Delta_a}{4\Delta_r} = 0.$$

[See article **47**, 3° and article **48**.]

2° The ellipse reduces to a line segment if $z_0 = 0$ or if $-\Delta_a/z_0^2\Delta_r$ is a real number not less than 1. It suffices, for example, to take $z_0 = a_0/r_0$, corresponding to $t = 0$. [See the remark of article **46** and use exercise 53, 2°.]

3° Write the equation of exercise 54 in the form (1).
For $z_0 = i$, $z = (3/2)e^{it_1} + (1/2)e^{-it_1}$.]

56. The equation

$$z = \frac{(1+i)(1+t)t}{1+2t+2t^2}$$

represents the segment of length $\sqrt{2}$, centered at the origin, and lying along the bisector of the angle formed by the positive axes. [See exercise 55, 2°.]

57. With the notation of article **59**, the equation

$$z = \frac{a_0 + 2a_1t + a_2t^2}{r_0 + 2r_1t + r_2t^2}, \qquad \Delta_r > 0$$

represents a line segment if

$$\frac{H}{2\Delta_r} = \frac{a_0}{r_0}$$

or if

$$\frac{(H^2 - 4\Delta_a\Delta_r)r_0^2}{(2a_0\Delta_r - Hr_0)^2}$$

is a real number not less than 1. [Take the origin at the center and use exercise 55, 2°.]

58. Generate, with the aid of two rotating vectors, the locus with equation

$$z = \frac{(1-i)t(1+t) - i}{(1+t)^2 + t^2}$$

and determine the nature of the locus. Construct the locus.

59. Prove, with the aid of the transformation

$$t_1 = \tan\frac{t}{2},$$

that if two certain trinomials are relatively prime, the equation

$$z = \frac{a_0 + 2a_1e^{it} + a_2e^{2it}}{b_0 + 2b_1e^{it} + b_2e^{2it}}$$

is that of a conic if b_1 is real and $\bar{b}_2 = b_0$, these numbers being able to be multiplied by a common imaginary factor.

The conic is an ellipse, a parabola, or a hyperbola according as $|b_0|$ is less than, equal to, or greater than $|b_1|$.

Find the conditions for the conic to be a circle [article **59**], an equilateral hyperbola [exercise 52], or a degenerate conic [exercises 48, 49, 51, 57].

VIII. UNICURSAL BICIRCULAR QUARTICS AND UNICURSAL CIRCULAR CUBICS

We first of all make the following remarks about the trinomials

$$A \equiv a_0 + 2a_1 t + a_2 t^2, \qquad B \equiv b_0 + 2b_1 t + b_2 t^2$$

in the real parameter t and whose coefficients are real or imaginary.

1° A *and* B *have no common zero and are therefore relatively prime* if and only if their eliminant

$$\Delta_{ab} = 4(a_0 b_1 - a_1 b_0).(a_1 b_2 - a_2 b_1) - (a_0 b_2 - a_2 b_0)^2$$

is not zero. If we set

$$m_0 = a_1 b_2 - a_2 b_1, \quad m_1 = a_2 b_0 - a_0 b_2, \quad m_2 = a_0 b_1 - a_1 b_0,$$

we have

$$\Delta_{ab} = 4m_0 m_2 - m_1^2.$$

2° *The zeros of* B *are real or conjugate imaginary* if and only if

$$\frac{b_0}{\bar{b}_0} = \frac{b_1}{\bar{b}_1} = \frac{b_2}{\bar{b}_2}, \tag{1}$$

for it is necessary and sufficient that b_0, b_1, b_2, and consequently also their conjugates, be proportional to (the same) three real numbers.

3° *For only one zero of* B *to be real,* it is necessary and sufficient that equations (1) do not hold and that the equations

$$B \equiv b_0 + 2b_1 t + b_2 t^2 = 0$$

$$\bar{B} \equiv \bar{b}_0 + 2\bar{b}_1 t + \bar{b}_2 t^2 = 0$$

be compatible, and therefore that we have

$$\Delta_{b\bar{b}} = 4(b_0\bar{b}_1 - b_1\bar{b}_0)(b_1\bar{b}_2 - b_2\bar{b}_1) - (b_0\bar{b}_2 - b_2\bar{b}_0)^2 = 0.$$

These conditions are necessary, for if we have (1) the two zeros of B will be real or conjugate imaginary, and if a real number is a root of an algebraic equation, it is also a root of the conjugate equation.

The conditions are sufficient, for if $b_0 = 0$ or $b_2 = 0$ we have the real zero 0 or ∞, and if $b_0 b_2 \neq 0$, the common zero is that of

$$\bar{b}_2 B - b_2 \bar{B} \equiv 2(b_1\bar{b}_2 - b_2\bar{b}_1)\, t + b_0\bar{b}_2 - b_2\bar{b}_0,$$

which is real.

4° From the preceding it follows that *for the zeros of* B *to be non-conjugate imaginary*, it is necessary and sufficient that we have

$$\Delta_{b\bar{b}} \neq 0,$$

for this inequality will not allow the equalities (1).

65. General equation. *If the trinomials* A, B *are relatively prime, the equation*

$$z = \frac{A}{B} = \frac{a_0 + 2a_1t + a_2t^2}{b_0 + 2b_1t + b_2t^2} \tag{2}$$

represents[1]

1° *a unicursal bicircular quartic if the zeros of* B *are non-conjugate imaginary, that is, if* $\Delta_{b\bar{b}} \neq 0$;

2° *a unicursal circular cubic if only one of these zeros is imaginary.*

We refer to the notation of article **54**.

1° Since the zeros of B are non-conjugate imaginary numbers, we have $n = 2$ and $\nu = 0$. Therefore

$$n - \nu = 2 \tag{3}$$

and the curve is bicircular (**56**). Also, since $m \leqq n$, we see that the curve is of order (**54**)

$$2n - \nu = 4.$$

Hence the curve is a unicursal bicircular quartic.

2° Since B contains only one imaginary zero, it is clear that $\nu = 0$ or 1 according as the other zero of B is infinite or a finite real number.

If $\nu = 0$, then $n = 1$ and

$$n - \nu = 1$$

and the curve is circular (**56**). We cannot have $m = 1$, for then A and B would each have infinity for a zero. Therefore $m = 2$ and $m > n$, whence the curve is of order (**54**)

$$m + n - \nu = 3.$$

Hence the curve is a unicursal circular cubic with an equation of the form

$$z = \frac{a_0 + 2a_1t + a_2t^2}{b_0 + 2b_1t} \tag{4}$$

with the condition

$$a_2 \neq 0, \quad \frac{b_0}{b_1} \text{ imaginary.}$$

[1] Recall that if the zeros of B are real or conjugate imaginary, the equation is that of a conic (§ VII).

If $\nu = 1$, then $n = 2$ and

$$n - \nu = 1$$

and the curve is circular (**56**). Also, since $m \leqq n$, we see that the curve is of order (**54**)

$$2n - \nu = 3.$$

Hence the curve is a unicursal circular cubic with an equation of the form

$$z = \frac{a_0 + 2a_1 t + a_2 t^2}{(b_0 + 2b_1 t)(r_0 + 2r_1 t)} \tag{5}$$

in which b_0/b_1, r_0/r_1 are respectively imaginary and real.

We can always reduce form (5) to form (4), and conversely. If we set

$$t = \frac{\alpha T + \beta}{\gamma T + \delta}, \quad \alpha\delta - \beta\gamma \neq 0, \quad \alpha, \beta, \gamma, \delta \text{ real}, \tag{6}$$

equation (5) becomes

$$z = \frac{a_0(\gamma T + \delta)^2 + 2\,a_1(\alpha T + \beta)(\gamma T + \delta) + a_2(\alpha T + \beta)^2}{[b_0(\gamma T + \delta) + 2\,b_1(\alpha T + \beta)]\,[r_0(\gamma T + \delta) + 2\,r_1(\alpha T + \beta)]}$$

and takes the form (4) if we can satisfy the relations

$$r_0\gamma + 2r_1\alpha = 0, \tag{7}$$

$$a_0\gamma^2 + 2a_1\alpha\gamma + a_2\alpha^2 \neq 0 \tag{8}$$

expressing that the coefficient of T in the second factor of the denominator is zero and that the coefficient of T^2 in the numerator is not zero. We can satisfy (7) by

$$\frac{\alpha}{\gamma} = -\frac{r_0}{2r_1}$$

inasmuch as r_0/r_1 is real, and (8) holds because the numerator and denominator of (5) are relatively prime.

Conversely, we reduce (4) to (5) with the aid of (6) where we assume $\gamma \neq 0$.

Corollary. *For the cubics, it suffices to consider equation* (2) *in which*

$$a_2 \neq 0, \quad b_2 = 0, \quad \frac{b_0}{b_1} \text{ imaginary}.$$

66. Double point. Theorem I. *The origin is a double point for the cubic or the quartic with equation* (2) *if the numbers* a_0, a_1, a_2 *are proportional to real numbers.*

If from (2) we obtain the cartesian coordinates x, y of a variable point of the curve as functions of t (54), then, for the origin to be a double point, it is necessary and sufficient that all lines through the origin intersect the curve in points of which two coincide with the origin, and hence that x and y vanish simultaneously for values of t which are real and distinct, real and equal, finite or infinite, or conjugate imaginary. Therefore it is necessary and sufficient that z vanish under the same conditions, that is, that the roots of

$$A \equiv a_0 + 2a_1t + a_2t^2 = 0$$

be real, finite or infinite, or be conjugate imaginary, whence the theorem.

Corollary. If a_0, a_1, a_2 *are real, then, according as*

$$\Delta_a \equiv a_0a_2 - a_1^2$$

is negative, zero, or positive, the two branches of the curve which pass through the double point D *have real and distinct tangents, real and coincident tangents, or conjugate imaginary tangents*, and D is called *a crunode, a cusp* or *spinode*, or *an acnode* or *isolated point.*

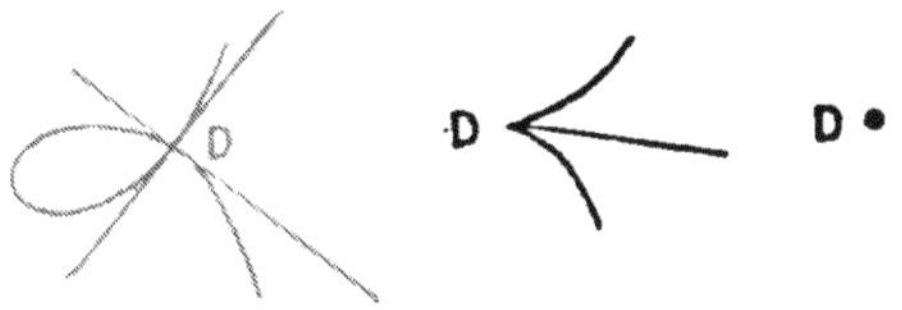

FIG. 41

Theorem II. *If equation* (2) *represents a quartic* C_4, *and if the numbers* m_0, m_1, m_2 *are not proportional to real numbers, then* C_4 *has a single proper double point* D *with affix*

$$d = \frac{a_0\bar{m}_0 + a_1\bar{m}_1 + a_2\bar{m}_2}{b_0\bar{m}_0 + b_1\bar{m}_1 + b_2\bar{m}_2} \tag{9}$$

given by the values of t *which are the roots of*

$$\begin{vmatrix} t^2 & 2t & 1 \\ m_2 & m_1 & m_0 \\ \bar{m}_2 & \bar{m}_1 & \bar{m}_0 \end{vmatrix} = 0. \tag{10}$$

D *is a crunode, cusp, or acnode according as the expression*

$$(m_0\bar{m}_1 - m_1\bar{m}_0)(m_1\bar{m}_2 - m_2\bar{m}_1) - (m_0\bar{m}_2 - m_2\bar{m}_0)^2 \tag{11}$$

is positive, zero, or negative.

The translation of axes to an arbitrary point D of affix d replaces equation (2) by

$$\zeta = \frac{a_0 - b_0 d + 2(a_1 - b_1 d)\, t + (a_2 - b_2 d)\, t^2}{b_0 + 2b_1 t + b_2 t^2} \tag{12}$$

and D is double for C_4 if there exist real numbers r_0, r_1, r_2 and a real or imaginary number k such that

$$b_0 d + k r_0 = a_0, \quad b_1 d + k r_1 = a_1, \quad b_2 d + k r_2 = a_2. \tag{13}$$

Since $\Delta_{b\bar{b}} \neq 0$, we do not have the equalities (1), the numbers b_0, b_1, b_2 are not proportional to real numbers, and the matrix

$$\left\| \begin{matrix} b_0 & b_1 & b_2 \\ r_0 & r_1 & r_2 \end{matrix} \right\| \tag{14}$$

is of rank 2. Consequently, in order that equations (13) in d and k be compatible, it is necessary and sufficient that we have

$$\begin{vmatrix} b_0 & r_0 & a_0 \\ b_1 & r_1 & a_1 \\ b_2 & r_2 & a_2 \end{vmatrix} = 0$$

or

$$m_0 r_0 + m_1 r_1 + m_2 r_2 = 0, \tag{15}$$

which implies

$$\bar{m}_0 r_0 + \bar{m}_1 r_1 + \bar{m}_2 r_2 = 0. \tag{16}$$

Since, by hypothesis, m_0, m_1, m_2 are not proportional to real numbers, the matrix

$$\left\| \begin{matrix} m_0 & m_1 & m_2 \\ \bar{m}_0 & \bar{m}_1 & \bar{m}_2 \end{matrix} \right\|$$

is of rank 2. All the real solutions of the system (15), (16) are therefore given by

$$r_0 = \lambda(m_1\bar{m}_2 - m_2\bar{m}_1), \quad r_1 = \lambda(m_2\bar{m}_0 - m_0\bar{m}_2), \quad r_2 = \lambda(m_0\bar{m}_1 - m_1\bar{m}_0), \tag{17}$$

where λ is an arbitrary pure imaginary number.

For these values, the second order determinants of (14) are, if we set

$$b_0\bar{m}_0 + b_1\bar{m}_1 + b_2\bar{m}_2 = \beta,$$

$$b_0 r_1 - b_1 r_0 = \lambda\, m_2\beta, \quad b_1 r_2 - b_2 r_1 = \lambda\, m_0\beta, \quad b_2 r_0 - b_0 r_2 = \lambda\, m_1\beta.$$

They are not all zero since (14) is of rank 2; consequently $\beta \neq 0$. Since m_0, m_1, m_2 are not all zero, we find from the first two of equations (13), if $m_2 \neq 0$. for example, the affix (9) of the unique double point.

The values t_1, t_2 of t which give (9) are such that

$$d = \frac{a_0 + 2a_1t_1 + a_2t_1^2}{b_0 + 2b_1t_1 + b_2t_1^2} = \frac{a_0 + 2a_1t_2 + a_2t_2^2}{b_0 + 2b_1t_2 + b_2t_2^2}$$

or, after performing some algebra,

$$2m_2 - m_1(t_1 + t_2) + 2m_0t_1t_2 = 0,$$

and consequently also, even if t_1, t_2 are conjugate imaginary,

$$2\bar{m}_2 - \bar{m}_1(t_1 + t_2) + 2\bar{m}_0t_1t_2 = 0.$$

Since the sought equation is of the form

$$t^2 - (t_1 + t_2)\, t + t_1t_2 = 0,$$

the elimination of $t_1 + t_2$ and t_1t_2 from these last three equations gives (10).

If we replace d in (12) by its value (9), equation (12) becomes

$$\zeta = -\frac{(m_1\bar{m}_2 - m_2\bar{m}_1) + 2(m_2\bar{m}_0 - m_0\bar{m}_2)\, t + (m_0\bar{m}_1 - m_1\bar{m}_0)\, t^2}{(b_0\bar{m}_0 + b_1\bar{m}_1 + b_2\bar{m}_2)\,(b_0 + 2b_1t + b_2t^2)},$$

whence, by the corollary of theorem I, we have the assertion concerning the sign of expression (11), for the numbers like $m_1\bar{m}_2 - m_2\bar{m}_1$ are pure imaginaries.

Corollary. *If the double point* D *is a cusp, its affix* d *is a root of the equation*

$$(b_0b_2 - b_1^2)d^2 - (a_0b_2 - 2a_1b_1 + a_2b_0)d + a_0a_2 - a_1^2 = 0 \qquad (18)$$

obtained by setting the discriminant of the trinomial in t in the numerator of (12) equal to zero.

In fact, the vanishing of (11) implies, by virtue of (17), the vanishing of $r_0r_2 - r_1^2$, which, because of (13), implies the vanishing of

$$(a_0 - b_0d)\,(a_2 - b_2d) - (a_1 - b_1d)^2.$$

Theorem III. *If equation* (2) *represents a quartic* C_4, *and if the numbers* m_0, m_1, m_2 *are proportional to real numbers* ρ_0, ρ_1, ρ_2, *the quartic degenerates into a circle* Γ *counted twice, and which has, according as* m_2, m_0, *or* m_1 *is different from zero, the equation*

$$z = \frac{a_0t_2 - a_1}{b_0t_2 - b_1}, \quad z = \frac{a_1t_0 - a_2}{b_1t_0 - b_2}, \quad z = \frac{a_2t_1 - a_0}{b_2t_1 - b_0},$$

t_2, t_0, t_1 *being real parameters.*

Because $\Delta_{ab} \neq 0$, at least one of the numbers m_0, m_1, m_2 is not zero. Since

$$\frac{m_0}{\bar{m}_0} = \frac{m_1}{\bar{m}_1} = \frac{m_2}{\bar{m}_2},$$

equations (15), (16) are equivalent and we can, if m_2, for example, is not zero, choose arbitrary real numbers r_0, r_1 and find r_2 by (15). We are going to show that, for r_0, r_1 not both zero, the first two of equations (13) are solvable for d and k, that is, that we have $b_0 r_1 - b_1 r_0 \neq 0$ or, what amounts to the same thing, $b_0 b_1 \neq 0$.

Because $\Delta_{b\bar{b}} \neq 0$, we have $b_0 \neq 0$. Otherwise, from the equations

$$a_1 b_2 - a_2 b_1 = \mu\rho_0, \quad a_2 b_0 - a_0 b_2 = \mu\rho_1, \quad a_0 b_1 - a_1 b_0 = \mu\rho_2,$$

we get

$$b_0\rho_0 + b_1\rho_1 + b_2\rho_2 = 0, \tag{19}$$

and consequently, if b_0 should be zero, b_0, b_1, b_2 would be proportional to real numbers and $\Delta_{b\bar{b}}$ would be zero.

Solving the first two of equations (13), we have

$$d = \frac{a_0 r_1 - a_1 r_0}{b_0 r_1 - b_1 r_0}.$$

Therefore C_4 has for double points all the points of the locus with equation

$$z = \frac{a_0 t_2 - a_1}{b_0 t_2 - b_1}. \tag{20}$$

By virtue of (19), b_1/b_0 is not real, since otherwise b_2/b_0 would also be real and we would have $\Delta_{b\bar{b}} = 0$. Consequently (41), the locus of double points is a circle and not a straight line.

Corollary. *If* $m_0 m_1 m_2 \neq 0$, *circle* Γ *contains the points with affixes* a_0/b_0, a_1/b_1, a_2/b_2. *The affix* ω *of its center and its radius* R *are* (42, 61)

$$\omega = \frac{a_1\bar{b}_2 - a_2\bar{b}_1}{b_1\bar{b}_2 - b_2\bar{b}_1} = \frac{a_2\bar{b}_0 - a_0\bar{b}_2}{b_2\bar{b}_0 - b_0\bar{b}_2} = \frac{a_0\bar{b}_1 - a_1\bar{b}_0}{b_0\bar{b}_1 - b_1\bar{b}_0},$$

$$\mathrm{R} = \left| \frac{m_0}{b_1\bar{b}_2 - b_2\bar{b}_1} \right| = \left| \frac{m_1}{b_2\bar{b}_0 - b_0\bar{b}_2} \right| = \left| \frac{m_2}{b_0\bar{b}_1 - b_1\bar{b}_0} \right|.$$

Theorem IV. *When the quartic* C_4 *degenerates into a double circle* Γ, *the point* Z *with affix* (2) *runs over the whole circle or over only an arc of the circle according as the number* $\rho_1^2 - 4\,\rho_0\rho_2$ *is negative or positive.*

Suppose $m_2 \neq 0$. The right members of equations (2) and (20) are equal if

$$t^2(m_1 t_2 + m_0) - 2m_2 t_2 t - m_2 = 0. \tag{21}$$

For a given real t_2, the two values of t are real if

$$\rho_2(\rho_2 t_2^2 + \rho_1 t_2 + \rho_0) \geqq 0. \tag{22}$$

Since $\Delta_{ab} = 4m_0 m_2 - m_1^2 \neq 0$, we also have $4\rho_0\rho_2 - \rho_1^2 \neq 0$. If $\rho_1^2 - 4\rho_0\rho_2 < 0$, relation (22) holds for all real values of t_2 because $\rho_2^2 > 0$, while if $\rho_1^2 - 4\rho_0\rho_2 > 0$, we must take $t_2 \leqq t'$ or $t_2 \geqq t''$, where we have designated the zeros of $\rho_2 t^2 + \rho_1 t + \rho_0$ by t', t'' $(t' < t'')$.

When $m_0 \neq 0$ or $m_1 \neq 0$, the analogues of (21) are

$$m_0 t_0 t^2 + 2m_0 t - (m_2 t_0 + m_1) = 0,$$
$$m_1 t^2 - 2(m_0 t_1 + m_2)t + m_1 t_1 = 0.$$

Corollary. *The two values* τ_1, τ_2 *of* t *which furnish a common point of* Γ *are related by*

$$2m_0\tau_1\tau_2 - m_1(\tau_1 + \tau_2) + 2m_2 = 0.$$

In fact, (21) yields the two relations

$$\tau_1 + \tau_2 = \frac{2m_2 t_2}{m_1 t_2 + m_0}, \quad \tau_1\tau_2 = \frac{-m_2}{m_1 t_2 + m_0},$$

from which it suffices to eliminate t_2.

Theorem V. *If equation* (2) *represents a cubic, then this cubic does not degenerate and it enjoys the properties stated in theorem II.*

By hypothesis we have

$$\Delta_{ab} = 4m_0 m_2 - m_1^2 \neq 0,$$
$$\Delta_{b\bar{b}} = 4(b_0\bar{b}_1 - b_1\bar{b}_0)(b_1\bar{b}_2 - b_2\bar{b}_1) - (b_0\bar{b}_2 - b_2\bar{b}_0)^2 = 0,$$

but the matrix

$$\left\| \begin{matrix} b_0 & b_1 & b_2 \\ \bar{b}_0 & \bar{b}_1 & \bar{b}_2 \end{matrix} \right\| \tag{23}$$

is of rank 2.

It suffices to show that m_0, m_1, m_2 cannot be proportional to real numbers ρ_0, ρ_1, ρ_2, because it is on this fact that rests, in theorem II, the existence of the unique proper double point.

We employ *reductio ad absurdum.* If we should have

$m_0 = a_1b_2 - a_2b_1 = \mu\rho_0,\ m_1 = a_2b_0 - a_0b_2 = \mu\rho_1,\ m_2 = a_0b_1 - a_1b_0 = \mu\rho_2,$

we would have

$$b_0\rho_0 + b_1\rho_1 + b_2\rho_2 = 0,$$

and hence also

$$\bar{b}_0\rho_0 + \bar{b}_1\rho_1 + \bar{b}_2\rho_2 = 0,$$

relations which, because of (23), give

$$\rho_0 = \nu(b_1\bar{b}_2 - b_2\bar{b}_1),\quad \rho_1 = \nu(b_2\bar{b}_0 - b_0\bar{b}_2),\quad \rho_2 = \nu(b_0\bar{b}_1 - b_1\bar{b}_0).$$

We would then have

$$\Delta_{ab} = \mu^2(4\rho_0\rho_2 - \rho_1^2) = \mu^2\nu^2\Delta_{b\bar{b}} = 0,$$

which is contrary to hypothesis.

67. Point construction of the cubic [1]

$$z = \frac{a_0 + 2a_1t + a_2t^2}{b_0 + 2b_1t}.$$

If we carry out the division, the equation becomes

$$z = \frac{4a_1b_1 - a_2b_0}{4b_1^2} + \frac{a_2}{2b_1}t + \frac{4a_0b_1^2 - 4a_1b_0b_1 + a_2b_0^2}{4b_1^2(b_0 + 2b_1t)}$$

and, by taking the origin at the point O_1 of affix $(4a_1b_1 - a_2b_0)/4b_1^2$, the equation takes the form

$$\zeta = at + \frac{b}{b_0 + 2b_1t}$$

when we set

$$a = \frac{a_2}{2b_1},\quad b = \frac{1}{4b_1^2}(4a_0b_1^2 - 4a_1b_0b_1 + a_2b_0^2).$$

Since the equations

$$z' = at,\quad z'' = \frac{b}{b_0 + 2b_1t}$$

are those of a straight line (d) and of a circle (c) passing through O_1, *each point Z of the cubic is given by the geometric sum of the vectors joining* O_1 *to the points of* (d) *and* (c) *which correspond to a common value of* t.

In the calibration of (d) and (c), point O_1 corresponds to 0 on (d) and to ∞ on (c).

[1] O. Bloch, *Die Ortskurven der graphischen Wechselstromtechnik nach einheitlicher Methode behandelt*, p. 58, Zürich, 1917.

The preceding construction shows that *the line* (d) *is the real asymptote of the cubic.* Its equation for the initial axes is

$$z = \frac{a_2}{2b_1} t + \frac{4a_1b_1 - a_2b_0}{4b_1^2}.$$

Construction of the double point D. The affix of the double point is given by (9), but *it is also obtainable as the point of intersection of the lines with equations*

$$z = \frac{a_0}{b_0} + \frac{a_2}{b_0} t', \qquad z = \frac{a_1}{b_1} + \frac{a_2}{b_1} t''. \tag{24}$$

In fact, equations (13), in which we assume $b_2 = 0$, give

$$\frac{a_0 - b_0 d}{a_2} = \frac{r_0}{r_2}, \quad \frac{a_1 - b_1 d}{a_2} = \frac{r_1}{r_2} \tag{25}$$

and if we denote the real numbers r_0/r_2, r_1/r_2 by $-t'$, $-t''$, equations (25) say that D is on the lines with equations (24). These lines are not coincident because, the quotient b_0/b_1 being imaginary, the lines are not parallel.

68. Inverse of a conic. *The transform of a non-circular conic* (C) *by an inversion of center* O *is a unicursal circular cubic or a unicursal bicircular quartic* (C_1) *with double point* O *according as* O *is or is not on* (C).

Choose axes with origin at O. The equation of any conic (C) is (**58**)

$$z = \frac{c_0 + 2c_1t + c_2t^2}{r_0 + 2r_1t + r_2t^2}, \tag{26}$$

where r_0, r_1, r_2 are real. Since in the inversion with center O and power p, the point with affix z has the point with affix $p/\bar{z}$ for inverse (**19**), the inverse curve (C_1) of (C) has for equation

$$z_1 = \frac{p}{\bar{z}} = \frac{p(r_0 + 2r_1t + r_2t^2)}{\bar{c}_0 + 2\bar{c}_1t + \bar{c}_2t^2} \tag{27}$$

and is therefore (**65, 66**) a unicursal circular cubic or a unicursal bicircular quartic with double point O.

(C_1) is a cubic if $c_2 = 0$ or if the trinomial $\bar{c}_0 + 2\bar{c}_1t + \bar{c}_2t^2$ has only one real zero r, which is then also a zero of $c_0 + 2c_1t + c_2t^2$. In the two cases (C) passes through O for $t = \infty$ and $t = r$ respectively.

Corollaries. 1° *The double point is an acnode, a cusp, or a crunode according as the conic* (C) *is an ellipse, a parabola, or a hyperbola* (**66**, corollary of theorem I).

The tangents to (C_1) *at* O *are parallel to the asymptotic directions of* (C). In fact, the inverse of a point at infinity on (C) is at O.

2° *The preceding statements are particular cases of general properties of inversion*, which we are going to recall.

An inversion with center O is a quadratic involutoric point transformation whose fundamental (or singular) points are O and the circular points I, J. [1] The transform of an algebraic curve (C) of order l which passes l_o, l_i, l_j times, respectively, through O, I, J, is a curve (C_1) of order $2l - l_o - l_i - l_j$ which passes $l - l_i - l_j$, $l - l_j - l_o$, $l - l_o - l_i$ times, respectively, through O, I, J. The tangents to (C_1) at O, I, J contain the points of intersection, other than O, I, J, of (C) with, respectively, the lines IJ, JO, OI ; two tangents at O to (C_1) coincide if (C) is tangent to IJ.

Therefore, if (C) is a non-circular conic, we have $l = 2$, $l_i = l_j = 0$, and (C_1) is a cubic or a quartic according as l_o has the value 1 or 0. We deduce from the preceding that *a necessary and sufficient condition for the circular points* I, J *to be cusps of the quartic* (C_1) *is that the center* O *be a focus of the conic* (C), for it is necessary and sufficient that the conic be tangent to the isotropic lines OI, OJ.

3° The converse of article **65** is true : *every unicursal circular cubic and every unicursal bicircular quartic can be represented by an equation of the form* (2).

We know that a cubic and a quartic are unicursal when they have, respectively, one and three double points. If the curves are circular, the quartic being bicircular, the unique double point of the cubic and the third double point of the quartic are necessarily in the finite part of the plane. An inversion having this double point O for center transforms the cubic or the quartic into a conic (C), for in the case of the cubic we have (see 2°)

$$l = 3, \quad l_o = 2, \quad l_i = 1, \quad l_j = 1, \quad 2l - l_o - l_i - l_j = 2,$$

and for the quartic we have

$$l = 4, \quad l_o = 2, \quad l_i = 2, \quad l_j = 2, \quad 2l - l_o - l_i - l_j = 2.$$

Since the equation of the conic (C) can always be put in the form (26) (**58**), the equation of the considered cubic or that of the considered quartic can always be given the form (27), therefore also the form (2).

---

[1] See, for example, K. Doehlemann, *Geometrische Transformationen*, II. Teil, p. 50 (Sammlung Schubert, XXVIII, Leipzig, 1908).

69. Limaçon of Pascal and cardioid. A limaçon of Pascal is a unicursal bicircular quartic whose circular double points are cusps. It is called a cardioid when the third double point D is also a cusp. In an inversion with center D, the limaçon then transforms (68, corollary 2°) into a conic with focus D. According as D is an acnode, a crunode, or a cusp, the conic is an ellipse, a hyperbola, or a parabola.

The quartic with equation

$$z = \frac{a_0 + 2a_1t + a_2t^2}{b_0 + 2b_1t + b_2t^2}$$

is a limaçon if

$$\Delta_b = b_0b_2 - b_1^2 = 0,$$

that is, if the denominator is a perfect square.

If we take the origin of axes at the double point D of affix d, the equation becomes

$$z_1 = \frac{a_0 - b_0d + 2(a_1 - b_1d)\,t + (a_2 - b_2d)\,t^2}{b_0 + 2b_1t + b_2t^2},$$

and we know (66, theorem I) that $a_0 - b_0d$, $a_1 - b_1d$, $a_2 - b_2d$ are proportional to real numbers.

The conic obtained from the quartic by the inversion with center D and power p has for equation

$$z_2 = \frac{p}{\bar{z}_1} = p\,\frac{\bar{b}_0 + 2\bar{b}_1t + \bar{b}_2t^2}{\bar{a}_0 - \bar{b}_0\bar{d} + 2(\bar{a}_1 - \bar{b}_1\bar{d})\,t + (\bar{a}_2 - \bar{b}_2\bar{d})\,t^2}.$$

In order that D be a focus, it is necessary and sufficient that we have (60, corollary)

$$\bar{b}_0\bar{b}_2 - \bar{b}_1^2 = 0 \quad \text{or} \quad b_0b_2 - b_1^2 = 0.$$

Corollary. *The quartic is a cardioid if* (66, theorem II)

$$b_0b_2 - b_1^2 = 0,$$

$$(m_0\bar{m}_1 - m_1\bar{m}_0)\,(m_1\bar{m}_2 - m_2\bar{m}_1) - (m_0\bar{m}_2 - m_2\bar{m}_0)^2 = 0.$$

70. Class of cubics and quartics considered. The number m of tangents which can be drawn from a point to a given plane algebraic curve is independent of the point and is called the *class* of the curve.

Plücker has shown that we have the relation

$$m = n(n-1) - 2d - 3r$$

connecting the order n, the class m, the number d of non-cuspidal double points, and the number r of cuspidal double points of the curve.

Consequently, *a unicursal circular cubic is of class* 4 *or* 3 *according as it is not or is cuspidal.*

A unicursal bicircular quartic is of class 6 *if none of its double points is a cusp, of class* 5 *if the proper double point is the only cusp, of class* 4 *or* 3 *according as the quartic is a limaçon which is not or is a cardioid.*

Remark. In the evaluation of the number m of tangents drawn from a point P of the curve, it is necessary to note that if P is simple, the tangent at this point counts for 2 in the number m; if P is double, the tangents at this point count for 4 or 3 according as P is not or is a cusp.

71. Foci. Theorem. *A unicursal circular cubic possesses a singular real focus and, according as it is or is not cuspidal, one or two ordinary real foci.*

A unicursal bicircular quartic which is not of the limaçon type possesses two singular real foci and, according as it is or is not cuspidal, one or two ordinary real foci. A limaçon has a singular real focus and an ordinary real focus; a cardioid has a singular real focus and no ordinary focus.

These properties follow from the definition of a focus (**57**) and from the class (**70**) of the curve. Thus, a non-cuspidal quartic has two tangents t_1, t_2 at the circular point I and the conjugate lines t'_1, t'_2 as tangents at the second circular point I′, whence the points $t_1t'_1$, $t_2t'_2$ are the two singular real foci. Moreover, since the class is 6, we can draw two tangents t_3, t_4 from I whose points of contact are not I, and the imaginary conjugate lines t'_3, t'_4 of t_3, t_4 are the tangents drawn from I′; the points $t_3t'_3$, $t_4t'_4$ are then the ordinary real foci.

Singular foci. If we take the origin of axes at the point F of affix ϕ, equation (2) becomes

$$\zeta = \frac{a_0 - b_0\phi + 2(a_1 - b_1\phi)\,t + (a_2 - b_2\phi)\,t^2}{b_0 + 2b_1t + b_2t^2}. \tag{28}$$

Point F is a singular focus if the numerator and the conjugate of the denominator have a common zero (**57**), and therefore if their eliminant is zero, or

$$4(\bar{b}_0\bar{b}_2 - \bar{b}_1^2)\,[(a_0 - b_0\phi)\,(a_2 - b_2\phi) - (a_1 - b_1\phi)^2] - [(a_0 - b_0\phi)\bar{b}_2 -$$
$$2(a_1 - b_1\phi)\,\bar{b}_1 + (a_2 - b_2\phi)\bar{b}_0]^2 = 0.$$

If we set, by analogy with what has been done for conics,

$$\Delta_a = a_0 a_2 - a_1^2, \quad \Delta_{\bar b} = \bar b_0 \bar b_2 - \bar b_1^2,$$

$$H_{ab} = a_0 b_2 - 2a_1 b_1 + a_2 b_0,$$

$H_{a\bar b}$ and $H_{b\bar a}$ having the obvious analogous meanings, the equation becomes

$$4\Delta_{\bar b}(\Delta_b \phi^2 - H_{ab}\phi + \Delta_a) - (H_{b\bar b}\phi - H_{a\bar b})^2 = 0 \tag{29}$$

or

$$(4\Delta_b \Delta_{\bar b} - H_{b\bar b}^2)\phi^2 - 2(2H_{ab}\Delta_{\bar b} - H_{a\bar b}H_{b\bar b})\phi + 4\Delta_a\Delta_{\bar b} - H_{a\bar b}^2 = 0. \tag{30}$$

For a cubic, the denominator in equation (2) has a real zero which is then a zero of the conjugate trinomial, and the eliminant $4\Delta_b\Delta_{\bar b} - H_{b\bar b}^2$ is zero. Equation (30) is of the first degree and the affix of the only singular focus of the cubic is

$$\frac{4\Delta_a\Delta_{\bar b} - H^2_{a\bar b}}{2(2H_{ab}\Delta_{\bar b} - H_{a\bar b}H_{b\bar b})}.$$

A quartic which is neither a limaçon nor a cardioid has two singular foci (see theorem); their affixes are the two roots of (30).

For the limaçon and the cardioid we have $\Delta_b = 0$ (**69**), hence also $\Delta_{\bar b} = 0$, and equation (29) gives the affix

$$\frac{H_{a\bar b}}{H_{b\bar b}}$$

of the only singular focus.

Ordinary foci. Their affixes are to be found among the values of ϕ which annul the discriminant of the numerator of (28) (**57**). We must therefore have

$$(a_0 - b_0\phi)(a_2 - b_2\phi) - (a_1 - b_1\phi)^2 = 0$$

or

$$\Delta_b\phi^2 - H_{ab}\phi + \Delta_a = 0. \tag{31}$$

This is the same equation as that for conics, except that the b's are imaginary (**60**).

We notice first of all that if the non-circular double point D is a cusp, its affix d is a root of equation (31), for this equation is none other than equation (18).

Consequently, *if the curve is not cuspidal at* D *or at the circular points, the roots of* (31) *are the affixes of the two ordinary foci.*

If the curve is cuspidal at D but is not a cardioid, the affix of its ordinary focus is

$$\frac{H_{ab}}{\Delta_b} - d \quad \text{or} \quad \frac{\Delta_a}{d\Delta_b}, \quad \text{if} \quad d \neq 0;$$

the affix of the ordinary focus of a limaçon is, since $\Delta_b = 0$ (69),

$$\frac{\Delta_a}{H_{ab}},$$

while for a cardioid this number is the affix of the cusp.

Corollaries. 1° *The equation*

$$z = \frac{(\alpha_0 + \alpha_1 t)^2}{b_0 + 2b_1 t + b_2 t^2}$$

is that of a unicursal circular cubic or unicursal bicircular quartic having an ordinary focus at the origin.

2° *The equation*

$$z = \left(\frac{\alpha_0 + \alpha_1 t}{\beta_0 + \beta_1 t} \right)^2 \tag{32}$$

is that of a limaçon of Pascal having its ordinary focus at the origin if α_0/α_1 is imaginary, and that of a cardioid having its cusp at the origin if α_0/α_1 is real (**66**, theorem I). *In both cases we assume β_0/β_1 is imaginary.*

72. Construction of the quartic. 1° **Limaçon and cardioid.** If we set

$$z_1 = \frac{\alpha_0 + \alpha_1 t}{\beta_0 + \beta_1 t}, \tag{33}$$

equation (32) becomes

$$z = z_1^2.$$

But equation (33) is that of a circle (c) whose center Ω has for affix (**61**)

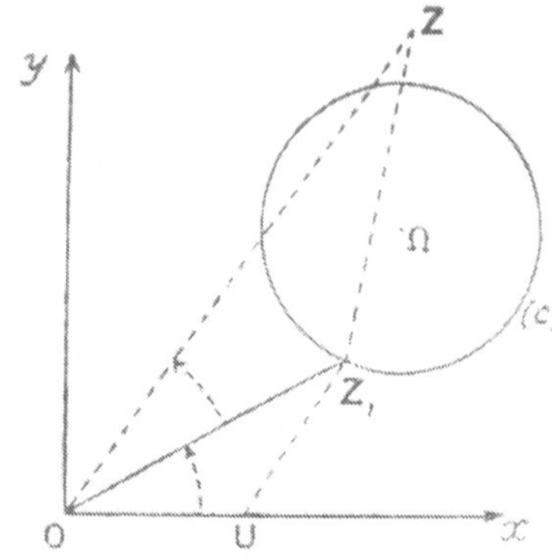

FIG. 42

$$\omega = \frac{\alpha_1 \bar{\beta}_0 - \alpha_0 \bar{\beta}_1}{\beta_1 \bar{\beta}_0 - \beta_0 \bar{\beta}_1} \tag{34}$$

and whose radius is

$$R = \left| \frac{\alpha_1 \beta_0 - \alpha_0 \beta_1}{\beta_1 \bar{\beta}_0 - \beta_0 \bar{\beta}_1} \right|.$$

If U, Z_1 are the point on Ox such that OU $= 1$ and a variable point on (c), then the triangle OZ_1Z

directly similar to triangle OUZ_1 gives a point Z of the curve (8) which corresponds to the same value of t as does Z_1.

We can also construct the curve, as an epicycloidal curve, with the aid of two rotating vectors.

In fact, the addition and subtraction of ω in the form (34) to and from the right member of (33) gives, if we set

$$a = \frac{\alpha_0\beta_1 - \alpha_1\beta_0}{\beta_1\bar{\beta}_0 - \beta_0\bar{\beta}_1},$$

$$z_1 = \omega + a\frac{\bar{\beta}_0 + t\bar{\beta}_1}{\beta_0 + t\beta_1}.$$

The coefficient of a has unit modulus and we can set

$$\frac{\bar{\beta}_0 + t\bar{\beta}_1}{\beta_0 + t\beta_1} = e^{i\tau}, \tag{35}$$

whence

$$z_1 = \omega + ae^{i\tau},$$

$$z = \omega^2 + 2a\omega e^{i\tau} + a^2e^{2i\tau}$$

and if we place the axes at the point of affix ω^2 the equation becomes

$$\zeta = 2a\omega e^{i\tau} + a^2e^{2i\tau}.$$

The vectors representing the numbers $2a\omega$, a^2 and rotating with angular speeds in the ratio 1 : 2 furnish the construction of the curve (50). Relation (35) between t and τ can, if we set

$$\beta_0 = r_0e^{i\theta_0}, \quad \beta_1 = r_1e^{i\theta_1},$$

be written as

$$t = -\frac{r_0 \sin\left(\theta_0 + \frac{\tau}{2}\right)}{r_1 \sin\left(\theta_1 + \frac{\tau}{2}\right)}.$$

2° **Other quartics.** If a_2 is not zero in equation (2), then, by carrying out the division of the two trinomials and by placing the origin at the point with affix a_2/b_2, the equation takes the form

$$\zeta = \frac{a'_0 + 2a'_1t}{b_0 + 2b_1t + b_2t^2}.$$

Designating the non-conjugate imaginary zeros of the denominator by p and q, we can write the preceding equation in the form

$$\zeta = \frac{l}{t-p} + \frac{m}{t-q},$$

where l and m are constants. Since the equations

$$\zeta_1 = \frac{l}{t-p}, \quad \zeta_2 = \frac{m}{t-q}$$

represent two circles passing through the origin ($t = \infty$), the quartic can be constructed pointwise by forming *the sum of the vectors issuing from the origin and terminating at the points of the circles which correspond to the same value of* t (also see article **64**).

3° **Double point.** *If the double point does not have one of the numbers* a_0/b_0, a_1/b_1, a_2/b_2, *relative to equation* (2), *as affix, then it belongs to the three straight lines or circles with equations*

$$z = \frac{a_1 - a_2 t}{b_1 - b_2 t}, \quad z = \frac{a_2 - a_0 t}{b_2 - b_0 t}, \quad z = \frac{a_0 - a_1 t}{b_0 - b_1 t}. \tag{36}$$

This follows from equations (13), from which we find, if we denote the quotient of the real numbers r_1, r_2 by t,

$$\frac{a_1 - b_1 d}{a_2 - b_2 d} = t, \qquad d = \frac{a_1 - a_2 t}{b_1 - b_2 t}.$$

The double point D with affix d is then on the straight line or the circle represented by the first of equations (36). We interpret the other two equations in a similar way.

Remark. For the graphical representation, and the establishment by classical analytic geometry of the properties of cubics and quartics, one may consult one of the following works :

GINO LORIA, *Spezielle algebraische und transszendente ebene Kurven*, 2 volumes, Berlin, 1910 ;

G. TEIXEIRA, *Tratado de las curvas especiales notables*, Madrid, 1905 ;

WIELEITNER, *Spezielle ebene Kurven* (Sammlungen Schubert und Göschen), Berlin ;

H. BROCARD and T. LEMOYNE, *Courbes géométriques remarquables*, Paris, 1919 (only one volume published, from *Abaque* to *Courbe auxiliaire* by alphabetical classification).

Exercises 60 through 71

60. 1° Study, and draw on quadrilruled paper, the curve with equation

$$z = \frac{t^2}{1 - it}.$$

[It is a cuspidal cubic with asymptote

$$z = 1 + it,$$

singular focus — 1/2, and ordinary focus 4.]

2° By associating the asymptote with a certain circle, show that the curve is a *cissoid of Diocles.*

61. Prove that the curve with equation

$$z = \frac{2(1 + it)}{1 + it + 2t^2}$$

is a cuspidal bicircular quartic which is not a limaçon of Pascal and whose cusp is at the origin O.

The inverse of the curve in the inversion of center O and power 1 is a cissoid of Diocles with singular focus at O.

Show that the quartic can be constructed pointwise with the aid of tangent circles having equations

$$z = \frac{8i}{3(t + i)}, \qquad z = \frac{4i}{3(i - t)},$$

by calibrating these on quadrilruled paper.

62. The inverse of the cissoid of exercise 60 in an inversion whose center is at the ordinary focus is a cardioid.

63. If the cubic or the quartic with equation

$$z = \frac{a_0 + 2a_1t + a_2t^2}{b_0 + 2b_1t + b_2t^2}$$

is such that the numerator vanishes for two real finite values t_k of t, the tangents at the origin contain the points with affixes

$$\frac{a_1 + a_2t_k}{b_0 + 2b_1t_k + b_2t_k^2}.$$

Examine the cases $a_2 = 0$ and $a_1 = a_2 = 0$. [See article **34.**]

64. Construct the curve with equation

$$z = \frac{1 - t^2}{1 - it}.$$

The tangents at the double point are perpendicular, the asymptote has the equation

$$z = -(1 + it),$$

the singular focus is at 1 on the curve, the ordinary foci are $2(-1 \pm \sqrt{2})$. Construct the curve with the aid of the asymptote and the circle with center O and radius 1. [The curve is a *right strophoid.*]

65. 1° The orthogonal projection P of a given point Z_0 on the tangent at a variable point of the parabola $\mathscr{P}$ with equation (article **62**)

$$z = it + t^2$$

describes the locus l having equation

$$z = \frac{z_0 - \bar{z}_0 + 2i(z_0 + \bar{z}_0)t - 2t^2}{2 + 4it}.$$

2° Locus l is a line (what line ?) when Z_0 is the focus of $\mathscr{P}$ and a cubic with double point Z_0 in all other cases.

3° When Z_0 is on the axis of $\mathscr{P}$, the cubic is called, in general, a *conchoid of Sluse.* Construct this cubic when Z_0 is the vertex of $\mathscr{P}$ (cissoid of Diocles), the foot of the directrix (right strophoid), the symmetric of the focus with respect to the directrix (trisectrix of Maclaurin).

66. If a bicircular quartic is composed of two coincident circles, equation (30) of article **71** has a double root and the affix of the center of the circle is

$$\frac{2H_{ab}\Delta_{\bar{b}} - H_{a\bar{b}}H_{b\bar{b}}^2 H}{4\Delta_b\Delta_{\bar{b}} - H_{b\bar{b}}^2}.$$

67. Show that the point with affix

$$z = \frac{2 + 4t + 2t^2}{2 - 3i - 2t - it^2}$$

describes an arc of the circle whose center has affix $i - 1$ and whose radius is $\sqrt{2}$, and that the endpoints of the arc have affixes

$$0 \quad \text{and} \quad \frac{4(3i - 1)}{5},$$

while the point with affix

$$z = \frac{1 + 2t + 2t^2}{1 + 2(1 + i)t - (2 + i)t^2}$$

describes the entire circle whose center has affix 1/2 and whose radius is 1/2. [Use **66**, theorems III and IV.]

68. Lemniscate of Booth. This name is applied to the quartics which are inverses of an ellipse or a hyperbola when the center of inversion is at the center of the conic. If the conic is an equilateral hyperbola, we have a *lemniscate of Bernoulli.*[1]

[1] G. LORIA, *Spezielle ebene Kurven*, vol. I, p. 134.

Show that the curves with equations

$$z = \frac{1 + t^2}{3 + 2it - 3t^2}, \qquad z = \frac{2i + t + t^2}{2 - it^2}$$

are, respectively, an elliptic lemniscate of Booth and a lemniscate of Bernoulli.

69. The quartic with equation

$$z = \frac{a_0 + 2a_1t + a_2t^2}{b_0 + 2b_1t + b_2t^2}$$

is a lemniscate of Booth if (using the notation of articles **66** and **71**)

$$\frac{a_0\bar{m}_0 + a_1\bar{m}_1 + a_2\bar{m}_2}{b_0\bar{m}_0 + b_1\bar{m}_1 + b_2\bar{m}_2} = \frac{\mathrm{H}_{ab}}{2\Delta_b}.$$

Deduce that the (inflexional) double point is the midpoint of the segment determined by the ordinary foci. Apply this to the lemniscate of exercise 68.

70. The equation

$$z = \frac{a_0 + 2a_1e^{it'} + a_2e^{2it'}}{b_0 + 2b_1e^{it'} + b_2e^{2it'}}$$

reduces to the form

$$z = \frac{\alpha_0 + 2\alpha_1t + \alpha_2t^2}{\beta_0 + 2\beta_1t + \beta_2t^2}$$

if we set $t = \tan t'/2$. [We have

$$e^{it'} = \frac{1 + it}{1 - it},$$

$$\alpha_0 = a_0 + 2a_1 + a_2, \qquad \alpha_1 = i(a_2 - a_0), \qquad \alpha_2 = -(a_0 - 2a_1 + a_2),$$

$$\beta_0 = b_0 + 2b_1 + b_2, \qquad \beta_1 = i(b_2 - b_0), \qquad \beta_2 = -(b_0 - 2b_1 + b_2).]$$

71. If we set

$$e^{it'} = \frac{\tau_1}{\tau_2}, \qquad t = \frac{t_1}{t_2},$$

the transformation executed in exercise 70 becomes the linear substitution

$$\tau_1 = it_1 + t_2, \qquad \tau_2 = -it_1 + t_2.$$

By considering the invariant of a binary quadratic form or the simultaneous invariant of two such forms, show (then verify directly) that

$$4\Delta_a = -\Delta_\alpha, \qquad 4\mathrm{H}_{ab} = -\mathrm{H}_{\alpha\beta}.$$

CHAPTER THREE

CIRCULAR TRANSFORMATIONS

I. GENERAL PROPERTIES OF THE HOMOGRAPHY

73. Definition. Let z, z' be the affixes of two points Z, Z' of the plane referred to two perpendicular axes Ox, Oy, and let α, β, γ, δ be given constants, real or imaginary.

Consider the equation

$$\boxed{\alpha z z' + \beta z + \gamma z' + \delta = 0} \tag{1}$$

which is bilinear in z and z.' By writing the equation in the form

$$z(\alpha z' + \beta) + \gamma z' + \delta = 0,$$

we see that if, having arbitrarily chosen z', we wish the equation to give one and only one value for z (∞ not excluded), it is necessary and sufficient that the equations

$$\alpha z' + \beta = 0, \quad \gamma z' + \delta = 0$$

be incompatible, or that

$$\boxed{\alpha\delta - \beta\gamma \neq 0.} \tag{2}$$

We then conclude, by writing the equation in the form

$$z'(\alpha z + \gamma) + \beta z + \delta = 0,$$

that to each value of z there corresponds one and only one value of z'.

Consequently, if the inequality (2) holds, equation (1) associates with each real point Z of the w plane one and only one point Z' of the superimposed w' plane, and conversely.

This particular *one-to-one transformation* (**13**) of the Gauss plane onto itself is called a *homographic transformation of the complex plane*, or, for a reason which will appear soon (**76, 77**), a *direct circular* *transformation* of the plane. It is also called a *Möbius transformation*, after the name of the German geometer who discovered it in 1853.[1]

Solved for z', equation (1) of the transformation becomes, by setting $-\beta = a$, $-\delta = b$, $\alpha = c$, $\gamma = d$,

$$z' = \frac{az + b}{cz + d}, \quad ad - bc \neq 0. \tag{3}$$

We say that z' is a *homographic function* of z.

74. Determination of the homography. *A homography is determined if to three distinct and arbitrarily chosen points* Z_1, Z_2, Z_3 *we assign, as correspondents or homologues, three distinct and arbitrarily chosen points* Z'_1, Z'_2, Z'_3.

If

$$z_1 \neq z_2 \neq z_3 \quad \text{and} \quad z'_1 \neq z'_2 \neq z'_3 \tag{4}$$

are the affixes of the given points, we must have the conditional system

$$\begin{aligned} \alpha z_1 z'_1 + \beta z_1 + \gamma z'_1 + \delta &= 0, \\ \alpha z_2 z'_2 + \beta z_2 + \gamma z'_2 + \delta &= 0, \\ \alpha z_3 z'_3 + \beta z_3 + \gamma z'_3 + \delta &= 0, \end{aligned} \tag{5}$$

in which the matrix

$$\begin{Vmatrix} z_1 z'_1 & z_1 & z'_1 & 1 \\ z_2 z'_2 & z_2 & z'_2 & 1 \\ z_3 z'_3 & z_3 & z'_3 & 1 \end{Vmatrix} \tag{6}$$

of the coefficients of the unknowns α, β, γ, δ is of rank 3, since otherwise we would have

$$\Delta_1 = \begin{vmatrix} z_1 z'_1 & z'_1 & 1 \\ z_2 z'_2 & z'_2 & 1 \\ z_3 z'_3 & z'_3 & 1 \end{vmatrix} = 0, \qquad \Delta_2 = \begin{vmatrix} z_1 & z'_1 & 1 \\ z_2 & z'_2 & 1 \\ z_3 & z'_3 & 1 \end{vmatrix} = 0$$

[1] Möbius, *Ueber eine neue Verwandtschaft zwischen ebenen Figuren*, Werke, 2, pp. 205-217.

and consequently also

$$\Delta_1 - z_1'\Delta_2 = \begin{vmatrix} 0 & z_1' & 1 \\ z_2(z_2' - z_1') & z_2' & 1 \\ z_3(z_3' - z_1') & z_3' & 1 \end{vmatrix} = \begin{vmatrix} 0 & 0 & 1 \\ z_2(z_2' - z_1') & z_2' - z_1' & 1 \\ z_3(z_3' - z_1') & z_3' - z_1' & 1 \end{vmatrix}$$

$$= (z_2' - z_1')(z_3' - z_1')(z_2 - z_3) = 0,$$

which is contrary to the hypothesis (4).

The numbers α, β, γ, δ are thus determined to within an arbitrary common factor, the homography is unique, and its equation

$$\begin{vmatrix} zz' & z & z' & 1 \\ z_1z_1' & z_1 & z_1' & 1 \\ z_2z_2' & z_2 & z_2' & 1 \\ z_3z_3' & z_3 & z_3' & 1 \end{vmatrix} = 0 \tag{7}$$

is obtained by insisting that equations (1) and (5) are satisfied by the values, which are not all zero, of α, β, γ, δ.

The above proof assumes that all the points are in the finite part of the plane. Three cases remain to be considered.

1° *A single point*, say Z_1, *is at infinity*. The first of equations (5) is replaced by

$$\alpha z_1' + \beta = 0 \tag{8}$$

and the matrix, corresponding to (6),

$$\begin{Vmatrix} z_1' & 1 & 0 & 0 \\ z_2z_2' & z_2 & z_2' & 1 \\ z_3z_3' & z_3 & z_3' & 1 \end{Vmatrix}$$

is of rank 3 as shown by the last three columns. The equation of the unique homography is then

$$\begin{vmatrix} zz' & z & z' & 1 \\ z_1' & 1 & 0 & 0 \\ z_2z_2' & z_2 & z_2' & 1 \\ z_3z_3' & z_3 & z_3' & 1 \end{vmatrix} = 0.$$

We see that this equation can be obtained from (7) by dividing the elements of the second line by z_1 and then letting z_1 approach infinity.

2° *A point of each triple is at infinity and the two points are not corresponding*, say the points Z_1, Z'_2. The first two of equations (5) are replaced by (8) and

$$\alpha z_2 + \gamma = 0.$$

The matrix

$$\left\| \begin{matrix} z'_1 & 1 & 0 & 0 \\ z_2 & 0 & 1 & 0 \\ z_3 z'_3 & z_3 & z'_3 & 1 \end{matrix} \right\|$$

is of rank 3 and the equation of the unique homography is

$$\begin{vmatrix} zz' & z & z' & 1 \\ z'_1 & 1 & 0 & 0 \\ z_2 & 0 & 1 & 0 \\ z_3 z'_3 & z_3 & z'_3 & 1 \end{vmatrix} = 0.$$

3° *Two corresponding points*, say Z_1, Z_1, *coincide with the point at infinity*. The conditional equations (5) are then

$$\alpha = 0,$$
$$\beta z_2 + \gamma z'_2 + \delta = 0,$$
$$\beta z_3 + \gamma z'_3 + \delta = 0,$$

and since the matrix

$$\left\| \begin{matrix} z_2 & z'_2 & 1 \\ z_3 & z'_3 & 1 \end{matrix} \right\|$$

is of rank 2, the unique homography has for equation

$$\begin{vmatrix} z & z' & 1 \\ z_2 & z'_2 & 1 \\ z_3 & z'_3 & 1 \end{vmatrix} = 0.$$

75. Invariance of anharmonic ratio. *The anharmonic ratio of any four points* Z_1, Z_2, Z_3, Z_4 *is equal to that of their correspondents* Z'_1, Z'_2, Z'_3, Z'_4 *in any homography of the complex plane.*

We must show that (25)

$$(z'_1 z'_2 z'_3 z'_4) = (z_1 z_2 z_3 z_4)$$

or, assuming the eight affixes are finite, that

$$\frac{z'_1 - z'_3}{z'_2 - z'_3} : \frac{z'_1 - z'_4}{z'_2 - z'_4} = \frac{z_1 - z_3}{z_2 - z_3} : \frac{z_1 - z_4}{z_2 - z_4}. \tag{9}$$

From equation (1) of the homography we obtain

$$z_1' - z_3' = \frac{(\alpha\delta - \beta\gamma)(z_1 - z_3)}{(\alpha z_1 + \gamma)(\alpha z_3 + \gamma)} \qquad (10)$$

and the analogous differences, whence we obtain equation (9). This equation can be verified also by starting with the second member.

Now suppose z_1' is infinite. This will be so if, when $\alpha \neq 0$, we have $z_1 = -\gamma/\alpha$, or, when $\alpha = 0$, we have $z_1 = \infty$. Since (27)

$$(\infty\, z_2' z_3' z_4') = (z_4' z_3' z_2' \infty) = \frac{z_4' - z_2'}{z_3' - z_2'}$$

we have, by expressions analogous to (10),

$$(\infty\, z_2' z_3' z_4') = \frac{z_4 - z_2}{\alpha z_4 + \gamma} : \frac{z_3 - z_2}{\alpha z_3 + \gamma}. \qquad (11)$$

In the first case, by replacing γ/α by $-z_1$, we obtain the desired relation; in the second case (11) becomes

$$(\infty\, z_2' z_3' z_4') - \frac{z_4 - z_2}{z_3 - z_2} = (z_4 z_3 z_2\, \infty) = (\infty\, z_2 z_3 z_4).$$

76. Circular transformation. The homographic transformation is said to be circular because of the following property.

Every curve of the set consisting of the straight lines and the circles of the plane is transformed into a curve of this set.

In fact, let l be any straight line or circle, Z_1, Z_2, Z_3 three fixed points and Z a moving point on l, and Z_1', Z_2', Z_3', Z' their correspondents in any homography of the complex plane.

The anharmonic ratio $(Z_1Z_2Z_3Z)$ is real (28); this is then also true of $(Z_1'Z_2'Z_3'Z')$ (75), and the point Z' thus describes the straight line or the circle l' passing through Z_1', Z_2', Z_3' (28).

Article **91** will determine when l' is a straight line.

77. Conservation of angles. *If two curves l, l_1 intersect in a point Z, whose affix is not $-d/c$, at an angle of algebraic value θ, then their transforms l', l_1' under the homography of equation* (3) *intersect in Z', the homologue of Z, at an angle $\theta + k\pi$, k an integer.*

We say that the homography conserves angles (to within $k\pi$) in both magnitude and sign, or, that the homography is a *directly conformal transformation.*

Let

$$z = f(t)$$

be the equation of l. The tangent at point Z is parallel to the vector $\overline{\mathrm{OT}}$ which represents dz/dt, and we can agree to orient the tangent positively in the sense of this vector.

The equation of l' is

$$z' = \frac{af(t) + b}{cf(t) + d}$$

and the tangent at the point Z′ corresponding to Z, which is assumed to be such that $cf(t) + d \neq 0$, is parallel to the vector $\mathrm{O}\tau'$ which represents

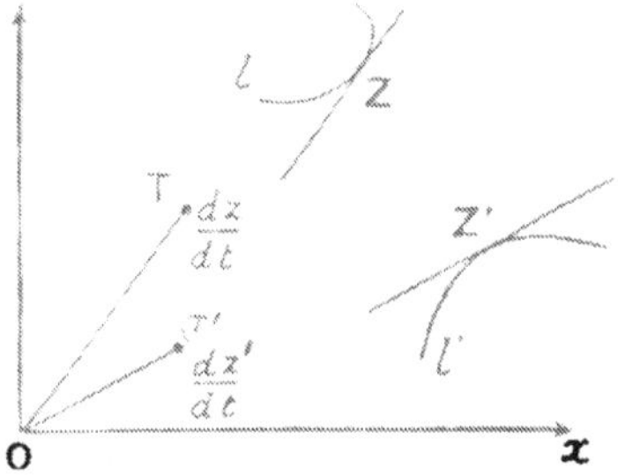

FIG. 43

$$\frac{dz'}{dt} = \frac{ad - bc}{(cz + d)^2}\,\frac{dz}{dt}.$$

If ϕ is the argument of $(ad - bc)/(cz + d)^2$, we then have

$$(\mathrm{O}x, \overline{\mathrm{O}\tau'}) = \phi + (\mathrm{O}x, \overline{\mathrm{OT}})$$

and, for the curves l_1, l'_1, by an obviously symmetric notation,

$$(\mathrm{O}x, \overline{\mathrm{O}\tau'_1}) = \phi + (\mathrm{O}x, \overline{\mathrm{OT}_1}).$$

From these relations we obtain by subtraction, to within $2k\pi$,

$$(\overline{\mathrm{O}\tau'}, \overline{\mathrm{O}\tau'_1}) = (\overline{\mathrm{OT}}, \overline{\mathrm{OT}_1}) = \theta.$$

If the positive sense of the tangent to l'_1 at Z′ is opposite to the sense of $\overline{\mathrm{O}\tau'_1}$, its angle with $\overline{\mathrm{O}\tau'}$ is then $\theta + (2k + 1)\pi$, which proves the theorem.

78. Product of two homographies. *The product of two homographies is a homography.*

Let

$$\alpha_1 z z_1 + \beta_1 z + \gamma_1 z_1 + \delta_1 = 0, \qquad \alpha_1\delta_1 - \beta_1\gamma_1 \neq 0 \qquad (12)$$

be the equation of a homography ω_1 in which to each point Z there corresponds a point Z_1, and

$$\alpha_2 z_1 z_2 + \beta_2 z_1 + \gamma_2 z_2 + \delta_2 = 0, \qquad \alpha_2\delta_2 - \beta_2\gamma_2 \neq 0 \qquad (13)$$

the equation of a homography ω_2 in which Z_1 corresponds to Z_2.

The product (21) $\omega = \omega_2\omega_1$ of these homographies is the transformation which permits us to pass from Z to Z_2. Its equation results from the elimination of z_1 from (12) and (13), and is

$$(\alpha_1\gamma_2 - \alpha_2\beta_1)zz_2 + (\alpha_1\delta_2 - \beta_1\beta_2)z + (\gamma_1\gamma_2 - \alpha_2\delta_1)z_2 + \gamma_1\delta_2 - \beta_2\delta_1 = 0, \quad (14)$$

a bilinear equation in z, z_2. Since

$$(\alpha_1\gamma_2 - \alpha_2\beta_1)(\gamma_1\delta_2 - \beta_2\delta_1) - (\alpha_1\delta_2 - \beta_1\beta_2)(\gamma_1\gamma_2 - \alpha_2\delta_1)$$
$$= (\alpha_1\delta_1 - \beta_1\gamma_1)(\alpha_2\delta_2 - \beta_2\gamma_2) \neq 0,$$

ω is a homography.

79. Circular group of the plane. A set of transformations constitutes a group if it satisfies the following two properties :

1° the product of any two transformations of the set is a transformation of the set ;

2° the inverse (13) of each transformation of the set is contained in the set.

The set of homographies

$$z' = \frac{az + b}{cz + d}, \qquad ad - bc \neq 0 \qquad (15)$$

possesses these two properties by virtue of article 78 and because the replacement of a by $-d$ and d by $-a$ converts equation (15) into the corresponding inverse transformation. We call this group the *circular group of the plane.*

80. Definitions. We call a point a **double point** of a homography if it coincides with its homologue.

A homography which is not the identity has at most two double points. In fact, there exists a single homography having three given double points (74), and the identity transformation possesses this property.

It is convenient to denote the Gauss plane by w or w' according as we consider it the set of points Z or that of their transforms Z′ under a homography.

The **limit point** L of the w plane for a homography ω is the point whose homologue L′ is at infinity in w'. The **limit point** M′ of the w' plane is the homologue of the point M ≡ L′ at infinity in the w plane. The points L, M′ are thus the homologues of the point at infinity in the homographies ω^{-1} and ω (13).

Exercises 72 through 74

72. 1° If a, b, c are the affixes of the points A, B, C, then the equation of the homography ω in which the points B, C, A correspond to the points A, B, C, and which is indicated by

$$\omega = \begin{pmatrix} A & B & C \\ B & C & A \end{pmatrix},$$

is

$$\alpha zz' + \beta z + \gamma z' + \delta = 0,$$

where

$$\alpha = ab + bc + ca - a^2 - b^2 - c^2, \qquad \beta = ac^2 + ba^2 + cb^2 - 3abc,$$
$$\gamma = ab^2 + bc^2 + ca^2 - 3abc, \qquad \delta = abc(a + b + c) - a^2b^2 - b^2c^2 - c^2a^2.$$

2° If A is the point at infinity in the plane, the equation of the homography is

$$zz' - bz - cz' + b^2 + c^2 - bc = 0.$$

73. 1° The homography ω^2, the square of the homography ω having equation

$$\alpha zz' + \beta z + \gamma z' + \delta = 0,$$

has for equation

$$\alpha(\gamma - \beta)zz'' + (\alpha\delta - \beta^2)z + (\gamma^2 - \alpha\delta)z'' + \delta(\gamma - \beta) = 0.$$

If ω is not the identity homography $z = z'$, ω^2 is the identity only if $\beta = \gamma$; ω is then an involution (articles **23** and **99**).

2° The equation of ω^3 is

$$\alpha(\beta^2 + \gamma^2 - \beta\gamma - \alpha\delta)zz''' + (\beta^3 + \alpha\gamma\delta - 2\alpha\beta\delta)z + (\gamma^3 + \alpha\beta\delta - 2\alpha\gamma\delta)z''' + \delta(\beta^2 + \gamma^2 - \beta\gamma - \alpha\delta) = 0,$$

and this homography is the identity if $\omega = 1$ or if,

$$\beta^2 + \gamma^2 = \beta\gamma + \alpha\delta.$$

3° ω^2 and ω^3 are involutions if we have, respectively,

$$\beta^2 + \gamma^2 - 2\alpha\delta = 0; \qquad \beta = \gamma \quad \text{or} \quad \beta^2 + \gamma^2 + \beta\gamma - 3\alpha\delta = 0,$$

and they cannot be involutions simultaneously.

4° The homography of exercise 72 is such that $\omega^3 = 1$. The verification is easy if A is at infinity; exercise 74 establishes the result more simply.

74. Cyclic homographies. A homography ω is cyclic with period n if $\omega^n = 1$, that is, if the homography ω^n transforms each point into itself. If $A_1, A_2, \ldots, A_{n-1}$, $A_n = A$ are the transforms of a point A by the homographies $\omega, \omega^2, \ldots, \omega^{n-1}, \omega^n$, we say that $A, A_1, A_2, \ldots, A_{n-1}$ constitute a *cyclic set* of ω. We shall consider the case $n = 3$. (The case $n = 2$ is that of the involution.)

1° The homography determined by

$$\omega = \begin{pmatrix} A & B & C \\ B & C & A \end{pmatrix}$$

possesses the cyclic triple A, B, C. Show that each point D is the start of a cyclic triple by establishing that if E is the homologue of D and F the homologue of E, then the homologue of F is D. [Use articles **75** and **29**.] Therefore ω is cyclic of period 3. Show this even more simply.

2° Obtain (exercises 72, 73) the identity

$$(ac^2 + ba^2 + cb^2 - 3abc)^2 + (ab^2 + bc^2 + ca^2 - 3abc)^2$$
$$- (ac^2 + ba^2 + cb^2 - 3abc)(ab^2 + bc^2 + ca^2 - 3abc)$$
$$- (ab + bc + ca - a^2 - b^2 - c^2)[abc(a + b + c) - a^2b^2 - b^2c^2 - c^2a^2] = 0.$$

3° If A′, B′, C′ are the harmonic conjugates of A, B, C, respectively, with respect to the pairs (B,C), (C,A), (A,B), the triple A′, B′, C′ is cyclic for ω.

4° We have

$$(AA'B'C') = (BB'C'A') = (CC'A'B') = -1.$$

Construct A′, B′, C′ (article 30). [See exercises 25; 26; 41, 8°.]

II. THE SIMILITUDE GROUP

81. Definition. The homography with equation

$$\beta z + \gamma z' + \delta = 0, \quad \beta\gamma \neq 0 \tag{1}$$

or

$$z' = \frac{az + b}{d}, \qquad ad \neq 0 \tag{2}$$

in which, α or c being zero, the zz' term does not figure, is called a *similitude*.

82. Properties. 1° *The point at infinity in the Gauss plane is a double point for every similitude. The two limit points coincide with the point at infinity* (**80**).

In fact, if z is infinite, equation (2) shows that z' is also infinite.

2° *Each straight line is transformed into a straight line and each circle is transformed into a circle.*

The curve with equation

$$z = \frac{pt + q}{rt + s}, \qquad ps - qr \neq 0$$

is a straight line or a circle according as r/s is real or imaginary (**41**).

Its transform by (2) has for equation

$$z' = \frac{(ap + br)\,t + aq + bs}{d(rt + s)}$$

and is a straight line or a circle according as r/s is real or imaginary, for in addition

$$(ap + br)s - (aq + bs)r = a(ps - qr) \neq 0.$$

3° *Each figure* (F) *transforms into a directly similar figure* (F′), whence the name of direct similitude has been given to the transformation, *and the ratio of similitude of* (F′) *to* (F) *is* $|\beta/\gamma|$ *or* $|a/d|$.

It suffices to consider two corresponding triangles $Z_1Z_2Z_3$, $Z'_1Z'_2Z'_3$. We have (**75, 82,** 1°)

$$(Z_1Z_2Z_3\,\infty) = (Z'_1Z'_2Z'_3\,\infty)$$

whence (**25**), by equating the moduli and then the arguments,

$$\left|\frac{Z_1Z_3}{Z_2Z_3}\right| = \left|\frac{Z'_1Z'_3}{Z'_2Z'_3}\right|, \qquad (\overline{Z_3Z_2}, \overline{Z_3Z_1}) = (\overline{Z'_3Z'_2}, \overline{Z'_3Z'_1}).$$

The two triangles are thus directly similar, and the ratio of similitude is

$$\left|\frac{Z'_1Z'_2}{Z_1Z_2}\right| = \left|\frac{z'_1 - z'_2}{z_1 - z_2}\right| = \left|\frac{az_1 + b - (az_2 + b)}{d(z_1 - z_2)}\right| = \left|\frac{a}{d}\right|.$$

4° *The direct similitudes constitute a group.*

In fact, the product of two similitudes is a homography which, also having the point at infinity as a double point, is a similitude; this fact also follows from equation (14) **78**, for if $\alpha_1 = \alpha_2 = 0$ we have $\alpha_1\gamma_2 - \alpha_2\beta_1 = 0$. In addition, the replacement of a by $-d$ and of d by $-a$ converts equation (2) into the corresponding inverse transformation, and this, we note, is a similitude.

The similitude group, being contained in the circular group, is a **subgroup** of the latter.

5° *Every direct similitude is the product of a rotation, a homothety having the same center, and a translation, where one or two of these component transformations may be wanting.*

Equation (2) follows, in fact, from the equations

$$z_1 = \frac{a}{d}\,z, \qquad z' = z_1 + \frac{b}{d}.$$

The first represents the identity transformation if $a = d$, a homothety with center at the origin if a/d is real but different from unity (**16**), a rotation about the origin if a/d is imaginary with unit modulus (**15**), the (permutable) product if a homothety and a rotation both centered at the origin if a/d is imaginary with non-unit modulus. The second represents a translation or the identity transformation.

83. Center of similitude. *A similitude which is neither the identity transformation nor a translation has a double point* Ω *in the finite part of the plane and called a center of similitude; the similitude is the permutable product of a rotation and a homothety both centered at* Ω*, one or the other of these two transformations perhaps being wanting.*

A translation has only the point at infinity fcr double point (82, 1°).

A point Ω of affix ω is a double point for the similitude with equation (2) if

$$\omega = \frac{a\omega + b}{d} \tag{3}$$

or

$$\omega(d - a) = b.$$

If $d = a$, $b = 0$, every point Ω is a double point, and (2) represents the identity transformation.

If $d = a$, $b \neq 0$, there is no finite point Ω, and (2) represents a translation.

If $d \neq a$, the only value of ω is

$$\omega = \frac{b}{d - a}.$$

For the origin at Ω we have

$$z = \zeta + \omega, \qquad z' = \zeta' + \omega$$

and equation (2) becomes

$$\zeta' + \omega = \frac{a(\zeta + \omega) + b}{d}$$

or, taking note of (3),

$$\zeta' = \frac{a}{d}\zeta$$

which completes the proof (82, 5°).

Construction. *If* Z_1, Z_1' *and* Z_2, Z_2' *are two pairs of corresponding points and if* P *is the point common to the lines* Z_1Z_2, $Z_1'Z_2'$, *the point*

Ω is the second point common to the circles $\gamma_1 = (Z_1Z_1'P)$, $\gamma_2 = (Z_2Z_2'P)$.

To each line d passing through Z_1, the similitude associates a line d' passing through Z_1' (**82**, 2°). If d rotates about Z_1, d' generates a pencil directly equal to that described by d (**77**); the angle (dd') is therefore constant and the point dd' lies on the circle γ_1 passing through Z_1, Z_1', P, for Z_1Z_2 and $Z_1'Z_2'$ are two lines d, d'. Since $Z_1\Omega$, $Z_1'\Omega$ are also two corresponding lines, Ω is on γ_1. A similar argument applied to the pair Z_2, Z_2' shows that Ω also lies on γ_2.

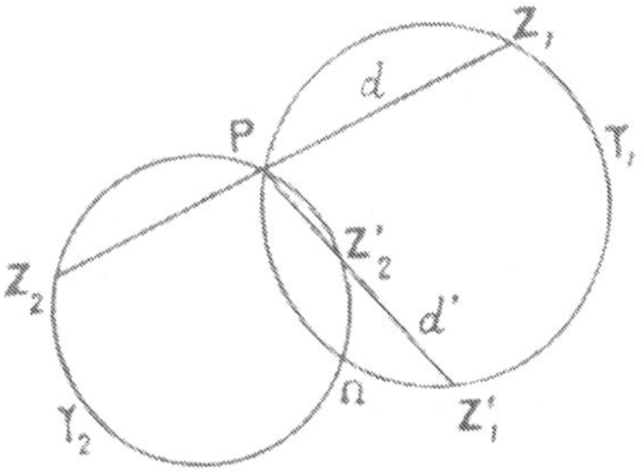

FIG. 44

If the lines Z_1Z_2, $Z_1'Z_2'$ are parallel, then, according as the vectors $\overline{Z_1Z_2}$, $\overline{Z_1'Z_2'}$ are not or are equal, the similitude is a homothety with center $\Omega = (Z_1Z_1', Z_2Z_2')$ or a translation.

84. Determination of a similitude. *A direct similitude is determined by two pairs of corresponding points.*

These two pairs, and the point at infinity in the Gauss plane considered as a double point, determine a homography (**74**, 3°). If the given points Z_1, Z_2 and their given homologues Z_1', Z_2' have z_1, z_2 and z_1', z_2' for affixes, the equation of the similitude is

$$\begin{vmatrix} z & z' & 1 \\ z_1 & z_1' & 1 \\ z_2 & z_2' & 1 \end{vmatrix} = 0.$$

Corollaries. 1° *A direct similitude is determined by its center and a pair of corresponding points.*

2° *A necessary and sufficient condition for two triangles* $Z_1Z_2Z_3$, $Z_1'Z_2'Z_3'$, *proper or degenerate, to be directly similar is that*

$$\begin{vmatrix} z_1 & z_1' & 1 \\ z_2 & z_2' & 1 \\ z_3 & z_3' & 1 \end{vmatrix} = 0.$$

3° *In order that a triangle* $Z_1Z_2Z_3$ *be equilateral,* it is necessary and sufficient that

$$\begin{vmatrix} z_1 & z_2 & 1 \\ z_2 & z_3 & 1 \\ z_3 & z_1 & 1 \end{vmatrix} = 0$$

or

$$z_1^2 + z_2^2 + z_3^2 = z_1 z_2 + z_2 z_3 + z_3 z_1$$

or again

$$(z_1 - z_2)^2 + (z_2 - z_3)^2 + (z_3 - z_1)^2 = 0.$$

In fact, it is necessary and sufficient that the triangles $Z_1Z_2Z_3$, $Z_2Z_3Z_1$ be directly similar.

85. Group of translations. *The translations of the plane form a group.*

The product of the translations with equations

$$z_1 = z + a_1, \qquad z_2 = z_1 + a_2$$

has for equation

$$z_2 = z + a_1 + a_2$$

and is a translation of vector equal to the sum of the vectors of the given translations.

Moreover, the set

$$z' = z + a$$

of translations contains the inverse of each transformation of the set.

86. Group of displacements. *The translations and the rotations of the plane form a group called the group of displacements.*

The similitude with equation (2) is a translation if the coefficient a/d of z has the value 1, and a rotation about the center of similitude if a/d is imaginary with modulus 1 (83). Each of the equations

$$z_1 = p_1 z + q_1, \qquad z_2 = p_2 z_1 + q_2 \tag{4}$$

then represents a translation or a rotation if $|p_1| = |p_2| = 1$, and the product of these similitudes having for equation

$$z_2 = p_2 p_1 z + p_2 q_1 + q_2 \tag{5}$$

is itself a translation or a rotation. Also, the inverse of $z_1 = p_1 z + q_1$ is $z_1 = z/p_1 - q_1/p_1$, which is again a translation or a rotation. It follows that the translations and the rotations form a group.

We call this group the group of displacements for the following reason. If a figure (F) is arbitrarily displaced in its plane so as finally to occupy a position (F′), then the figures (F), (F′) are directly equal. That is, (F′) corresponds to (F) under a direct similitude whose ratio of similitude is 1. This similitude then has an equation of the form

$$z' = pz + q, \qquad |p| = 1$$

and is a translation or a rotation capable, just as the considered displacement, of carrying (F) into coincidence with (F′).

Corollaries. 1° *Every transformation of this group conserves the magnitude and the sense of each angle, as well as the length of each segment.*

2° *The rotations do not constitute a group, but rotations having the same center do constitute a group.*

In fact, equations (4) represent rotations if

$$p_1 = e^{i\theta_1}, \qquad p_2 = e^{i\theta_2}$$

and (5) represents a translation if

$$e^{i(\theta_1+\theta_2)} = 1 \quad \text{or} \quad \theta_1 + \theta_2 = 2k\pi, \ (k \text{ an integer}).$$

On the other hand, if the rotations have the same center, say the origin, we have $q_1 = q_2 = 0$ and (5) represents a rotation.

87. Group of translations and homotheties. The equation

$$z' = pz + q$$

represents a translation if $p = 1$, and a homothety if p is real but different from 0 and 1 (83). If each of the equations (4) represents a translation or a homothety, then so also does equation (5). Moreover, the inverse of a translation or a homothety is another translation or homothety. The translations and the homotheties therefore form a group.

Corollaries. 1° *Every transformation of this group conserves the magnitude and the sense of each angle, as well as the direction of each segment, and changes the length of each segment in a constant ratio.*

2° *The homotheties do not constitute a group, but the homotheties having the same center do constitute a group.*

The product of two homotheties not having the same center is a translation if the ratios of the homotheties are reciprocals of one another, otherwise it is a homothety whose center is on the line which joins the centers of the given homotheties.

Suppose the center of the first homothety, with equation (4), is at the origin; then $q_1 = 0$. If $q_2 \neq 0$, (5) represents a translation if $p_1p_2 = 1$, a homothety if $p_1p_2 \neq 1$; when $q_2 = 0$, (5) represents a homothety or the identity transformation. If (4) and (5) represent homotheties, the affixes of their centers, with $q_1 = 0$ and $q_2 \neq 0$, are

$$0, \quad \frac{q_2}{1-p_2}, \quad \frac{q_2}{1-p_1p_2}$$

and are thus collinear.

88. Permutable similitudes. *Two direct similitudes are permutable* (22) *if they are two translations or if they have the same center.*

If we reverse the order of the similitudes ω_1, ω_2 characterized by the coefficients p_1, q_1 and p_2, q_2 of equations (4), these equations are replaced by

$$z_1' = p_2z + q_2, \qquad z_2' = p_1z_1' + q_1$$

and the equation of the product $\omega_1\omega_2$ is

$$z_2' = p_1p_2z + p_1q_2 + q_1. \tag{6}$$

We will have $\omega_2\omega_1 = \omega_1\omega_2$ if the second members of equations (5) and (6) are equal, and hence if

$$q_1(1-p_2) = q_2(1-p_1). \tag{7}$$

When $p_1 = 1$, so that ω_1 is a translation, it is necessary that $p_2 = 1$ or $q_1 = 0$, that is, that ω_2 be a translation or ω_1 be the identity transformation. The last case affirms that the identity transformation is permutable with every similitude, which is obvious.

When p_1 and p_2 are different from 1, equation (7) states that

$$\frac{q_1}{1-p_1} = \frac{q_2}{1-p_2},$$

and hence that the centers of similitude of ω_1 and ω_2 coincide (83).

89. Involutoric similitude. *The only direct involutoric similitude is the symmetry with respect to a point.*

The direct similitude with equation

$$z' = pz + q \tag{8}$$

is involutoric if (23) it coincides with its inverse with equation

$$z = pz' + q,$$

and therefore if for all z

$$z = p^2z + pq + q$$

so that

$$p^2 = 1, \quad q(p+1) = 0$$

If $p = 1$, it is necessary that $q = 0$, and equation (8) represents the identity transformation. If $p = -1$, we have

$$z' = -z + q$$

which represents (**15**, **16**) the symmetry with respect to the point of affix $q/2$.

90. Application. We shall examine some properties of the following figure which is encountered in electrical studies: *being given a triangle* ABC, *consider the triples of directly similar triangles* BCA′, CAB′, ABC′.

One of these triples is determined when A′ is chosen, and we have the three direct similitudes

$$\alpha = (\text{CAB}', \text{ABC}'), \quad \beta = (\text{ABC}', \text{BCA}'), \quad \gamma = (\text{BCA}', \text{CAB}').$$

If we designate by a the affix of A in some rectangular cartesian system, the equation of α in running complex coordinates b', c' is (**84**)

$$\begin{vmatrix} c & a & 1 \\ a & b & 1 \\ b' & c' & 1 \end{vmatrix} = 0$$

or

$$b'(a - b) + c'(a - c) + bc - a^2 = 0. \qquad (9)$$

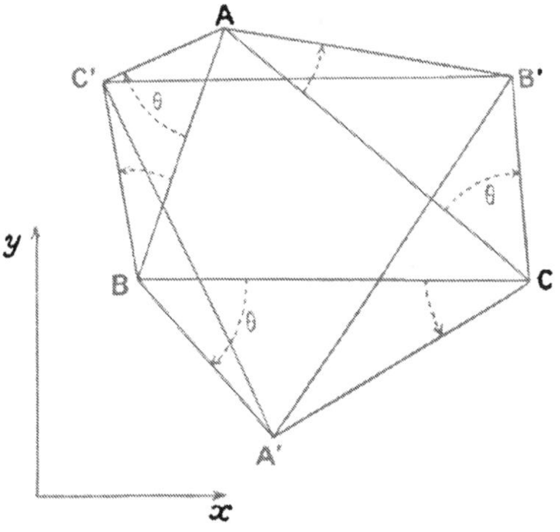

Fig. 45

Similarly, the equations of β, γ are

$$c'(b - c) + a'(b - a) + ca - b^2 = 0,$$
$$a'(c - a) + b'(c - b) + ab - c^2 = 0.$$

1° *The centers of similitude* S_a, S_b, S_c *of* α, β, γ *are, respectively, the orthogonal projections of the circumcenter of triangle* ABC *on the symmedians through* A, B, C.[1]

Equation (9) gives

$$s_a = \frac{bc - a^2}{b + c - 2a}.$$

[1] S_a, S_b, S_c are, in the geometry of the triangle, the vertices of the second Brocard triangle.

If the origin is assumed at A, then $a = 0$ and we have

$$\frac{1}{s_a} = \frac{1}{b} + \frac{1}{c},$$

whence the indicated construction for S_a (31, 2°).

2° *Point* S_a *is common to the circle passing through* C *and tangent to* AB *at* A *and the circle passing through* B *and tangent to* AC *at* A.

This construction is that of article 83 utilizing the two pairs of equal pencils

$$C(AS_a...) \,\overline{\wedge}\, A(BS_a...), \qquad A(CS_a...) \,\overline{\wedge}\, B(AS_a...)$$

if we note that to the ray in the first pencil joining the centers C and A, there must correspond in the second pencil the tangent AB at the center A.

3° *The centroid of triangle* A′B′C′ *is fixed and coincides with the centroid* G *of triangle* ABC.

Point A′ is determined if we know

$$\left|\frac{BA'}{BC}\right| \quad \text{and} \quad (\overline{BC}, \overline{BA'}) = \theta.$$

If we set

$$\left|\frac{BA'}{BC}\right| e^{i\theta} = \lambda$$

we then have

$$\frac{a' - b}{c - b} = \frac{b' - c}{a - c} = \frac{c' - a}{b - a} = \lambda$$

and consequently

$$a' = b(1 - \lambda) + c\lambda, \tag{10}$$

$$b' = c(1 - \lambda) + a\lambda, \tag{11}$$

$$c' = a(1 - \lambda) + b\lambda. \tag{12}$$

The vertices of triangle A′B′C′ are thus fixed by means of a common complex parameter λ.

Addition of equations (10), (11), (12), member by member, gives

$$a' + b' + c' = a + b + c,$$

which establishes the property (38).

4° *If* ABC *is equilateral, so also is* A′B′C′. If ABC *is not equilateral, the only two equilateral triangles* A′B′C′ *have their vertices at the centroids of the equilateral triangles constructed on* BC, CA, AB, *all exterior to triangle* ABC *or all interior to triangle* ABC.

A'B'C' is equilateral if (84)

$$\begin{vmatrix} b(1-\lambda)+c\lambda & c(1-\lambda)+a\lambda & 1 \\ c(1-\lambda)+a\lambda & a(1-\lambda)+b\lambda & 1 \\ a(1-\lambda)+b\lambda & b(1-\lambda)+c\lambda & 1 \end{vmatrix} = 0$$

or, by breaking up the determinant into a sum of four determinants, if

$$\begin{vmatrix} a & b & 1 \\ b & c & 1 \\ c & a & 1 \end{vmatrix} (3\lambda^2 - 3\lambda + 1) = 0.$$

This relation holds for all values of λ if ABC is equilateral, and in the contrary case if

$$\lambda = \frac{3 \pm i\sqrt{3}}{6} = \frac{\sqrt{3}}{3} e^{\pm i\pi/6},$$

whence the theorem.

5° *The points* A', B', C' *are collinear if they respectively describe the circumcircles of triangles* GS_bS_c, GS_cS_a, GS_aS_b, *the line* A'B'C' *passing through* G.[1]

The collinearity of A', B', C' holds if (36, 4°), t being a real parameter, we have

$$b' - a' = t(c' - a')$$

or, by taking note of (10), (11), (12), if

$$\lambda = \frac{(a-b)\,t + (b-c)}{(a+c-2b)\,t + a + b - 2c}.$$

If we substitute this value in (10), we obtain for a parametric equation of the locus of A'

$$a' = \frac{(ac-b^2)\,t + ab - c^2}{(a+c-2b)\,t + a + b - 2c}.$$

Since we have

$$\begin{vmatrix} ac-b^2 & a+c-2b \\ ab-c^2 & a+b-2c \end{vmatrix} = \begin{vmatrix} ac-b^2 & a+c-2b \\ (b-c)(a+b+c) & 3(b-c) \end{vmatrix} =$$

$$(b-c)(a^2+b^2+c^2-ab-bc-ca) \neq 0$$

[1] These three circles are the McCay circles of triangle ABC. One can find further properties in our study, *Triangles associés à trois figures semblables*, *Mathesis*, 1931, pp. 181-186.

if ABC is not equilateral, and since, B_1 and C_1 being the midpoints of the sides CA and AB, the quotient

$$\frac{a+c-2b}{a+b-2c} = \frac{\dfrac{a+c}{2}-b}{\dfrac{a+b}{2}-c} = \left|\frac{BB_1}{CC_1}\right| e^{i(\overline{CC_1},\,\overline{BB_1})}$$

is imaginary, the locus of A′ is a circle (41). The values $-1, 0, \infty$ for t give the points G, S_c, S_b. We similarly obtain the loci of B′ and C′. The line A′B′C′ passes through G by virtue of 3°.

Exercises 75 through 83

75. Given two circles having equations

$$z\bar{z} = R_1^2, \qquad z\bar{z} - az - a\bar{z} + a^2 - R_2^2 = 0,$$

there exists an infinity of direct similitudes

$$\zeta = \alpha\zeta' + \beta$$

which transform the first into the second. They are such that α is any number of modulus R_1/R_2 and $\beta = -a\alpha$.

If $R_1 = R_2$, these similitudes are rotations centered on the radical axis of the circles. If $R_1 \neq R_2$, the center of similitude describes the circle having equation

$$\left(\frac{R_2^2}{R_1^2} - 1\right)z\bar{z} + a(z + \bar{z} - a) = 0.$$

76. Find the affixes of, and construct, the pairs of corresponding points of a given direct similitude which, with a given point, are the vertices of an equilateral triangle.

77. On each pair of corresponding points Z, Z′ in a direct similitude is constructed the triangle ZZ′Z″ directly similar to a given triangle ABC. Show that the figure consisting of the points Z″ is directly similar to that consisting of the points Z and to that consisting of the points Z′.

78. Determine the locus of pairs of corresponding points of a direct similitude which subtend a right angle at a given point.

79. Construct a triangle knowing the vertices of the equilateral triangles constructed exteriorly (interiorly) on its sides.

Generalize : construct n coplanar points A_k ($k = 1, 2, ..., n$) knowing the vertices B_k of the triangles $A_kA_{k+1}B_k$ directly similar to a given proper or degenerate triangle. [See *Mathesis*, 1950, p. 262.]

80. Two points A_1, A_2 describe two coplanar circles (O_1) and (O_2) with constant and opposite angular velocities. If K is a fixed point, determine the locus of the vertex P of the triangle KA_2P directly similar to the triangle KO_1A_1. [It is an ellipse, a circle, or a line segment. See *Mathesis*, 1950, p. 353.]

81. If BCA_1A_2, CAB_1B_2, ABC_1C_2 are directly similar figures, and if A_1B_2P, B_1C_2Q, C_1A_2R also are directly similar figures, then the triangles ABC, PQR have the same centroid. [See *Mathesis*, 1953, p. 347.]

82. 1° In the application of article **90**, we can construct three points A'', B'', C'' (collinear or vertices of a triangle) satisfying the vector relations

$$\overline{B''C''} = \overline{AA'}, \qquad \overline{C''A''} = \overline{BB'}, \qquad \overline{A''B''} = \overline{CC'}.$$

2° If the triangle ABC is equilateral, the figures A'B'C', A''B''C'' are directly similar no matter what point be chosen for A'.

3° If ABC is not equilateral, these figures are directly similar only if A'BC is a right isosceles triangle with right angle at A'; the figures are then actually equal with corresponding segments perpendicular to one another.

[For the collinearity of A'', B'', C'', which occurs if AA', BB', CC' are parallel, see *Mathesis*, 1931, pp. 181-186.]

83. A direct similitude transforms a unicursal algebraic curve of order n and k-circular into a curve of the same type; it transforms the centroid, the orthocenter, the circumcenter, and the isodynamic centers of any triangle into these same points in the corresponding triangle.

III. NON-SIMILITUDE HOMOGRAPHY

91. Limit points. The homography with equation

$$z' = \frac{az + b}{cz + d}, \quad ad - bc \neq 0 \tag{1}$$

possesses two limit points L, M' (80) with affixes

$$l = -\frac{d}{c}, \quad m' = \frac{a}{c}. \tag{2}$$

Theorem I. *The equation of the homography can be written as*

$$(z - l)(z' - m') = \frac{bc - ad}{c^2}. \tag{3}$$

In fact, equation (1) or

$$czz' + dz' - az - b = 0$$

can, by taking note of (2), be written as

$$zz' - lz' - m'z - \frac{b}{c} = 0$$

or

$$(z-l)\,z'-(z-l)\,m'-lm'-\frac{b}{c}=0\ ;$$

which is easily put in the form (3).

Theorem II. *For a straight line or a circle* (C) *of the* w *plane to transform into a straight line* (C′) *of the* w′ *plane, it is necessary and sufficient that* (C) *contain the limit point* L *of* w.

It is necessary, for if (C′) is a straight line it must contain the point L′ at infinity in w', and then (C) contains L.

It is sufficient, for if (C) contains L, the transform (C′) contains L′ and is therefore a straight line.

Corollary. *The pencil of lines having vertex* L *is the only pencil which transforms into a pencil of lines, and this pencil has* M′ *for vertex.*

Theorem III. *The corresponding pencils of lines having their vertices at the limit points* L, M′ *are* **inversely** *equal, and if* Z, Z′ *are any two corresponding points, the product*

$$|\ \mathrm{LZ.M'Z'}\ |$$

is a constant.

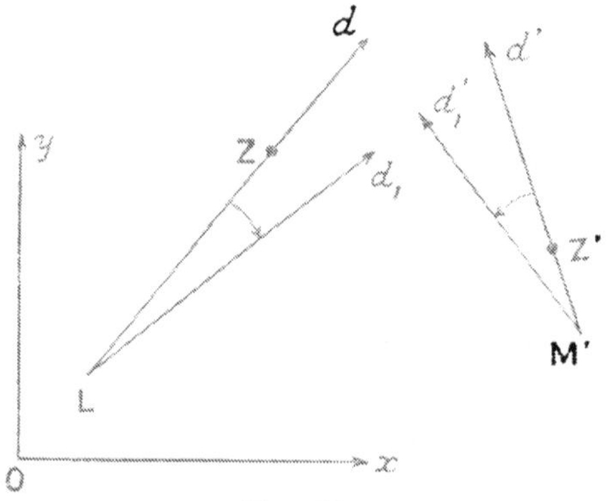

FIG. 46

On the lines LZ, M′Z′ place axes d, d' whose positive senses are those of L toward Z and M′ toward Z′. If

$$\phi=(xd),\qquad \phi'=(xd')$$

and if θ, r are the argument and the modulus of

$$(bc-ad)/c^2,$$

we have

$$z-l=\mathrm{LZ}.e^{i\phi},\quad z'-m'=\mathrm{M'Z'}.e^{i\phi'}.$$

From equation (3), which can be written as

$$\mathrm{LZ.M'Z'}.e^{i(\phi+\phi')}=re^{i\theta}$$

we obtain

$$\mathrm{LZ\,.\,M'Z'}=r=\left|\frac{bc-ad}{c^2}\right|=\text{constant}$$

and, to within an integral multiple of 2π,

$$\phi+\phi'=\theta.$$

If d_1, d_1' is a second pair of corresponding axes similar to d, d', we have

$$\phi_1 + \phi_1' = \theta,$$

whence

$$\phi_1 - \phi = -(\phi_1' - \phi'), \quad (xd_1) - (xd) = -[(xd_1') - (xd')],$$

$$(dd_1) = -(d'd_1'),$$

which establishes the inverse equality of the pencils.

Remark. *The angle of the two axes intersecting in the limit point* L *is conserved in magnitude but not in sense* (see article **77**).

92. Double points. Theorem I. *If a homography of the complex plane is not a similitude, it possesses two double points* E *and* F *in the finite part of the plane and the midpoint of segment* EF *coincides with the midpoint of the segment determined by the limit points* L *and* M′.

When the double points coincide, the homography is said to be **parabolic.**

The affixes e, f of the double points (80) are the roots of the equation

$$cz^2 - (a-d)z - b = 0 \tag{4}$$

obtained by setting $z' = z$ in (1), whence

$$e = \frac{1}{2c}[a - d + \sqrt{(a-d)^2 + 4bc}], \quad f = \frac{1}{2c}[a - d - \sqrt{(a-d)^2 + 4bc}].$$

The homography is parabolic if

$$(a-d)^2 + 4\,bc = 0.$$

The segments EF, LM′ have the same midpoint, for from (4) and (2) we obtain

$$e + f = \frac{a-d}{c} = l + m'.$$

Theorem II. *If the homography is not parabolic, the anharmonic ratio formed by the double points and any two corresponding points is a real or imaginary constant* λ, *different from* 0, 1, ∞.

This property applies also to a similitude which is not a translation.

For two pairs Z, Z′ and Z_1, Z_1' of corresponding points we have (75)

$$(efzz_1) = (efz'z_1')$$

or

$$\frac{e-z}{f-z} : \frac{e-z_1}{f-z_1} = \frac{e-z'}{f-z'} : \frac{e-z_1'}{f-z_1},$$

$$\frac{e-z}{f-z} : \frac{e-z'}{f-z'} = \frac{e-z_1}{f-z_1} : \frac{e-z_1'}{f-z_1'},$$

$$(efzz') = (efz_1z_1') = \lambda.$$

By taking Z $=$ L, Z$'$ $=$ L$'$ $= \infty$, we have (91)

$$\lambda = (efl\infty) = \frac{e-l}{f-l} = \frac{ce+d}{cf+d} = \frac{a+d+\sqrt{(a-d)^2+4bc}}{a+d-\sqrt{(a-d)^2+4bc}}. \quad (5)$$

In the case of the similitude having equation (1) with $c = 0$ and $a \neq d$, point E is the center of similitude and point F is at infinity, and we have

$$e = \frac{b}{d-a},$$

$$(e\infty zz') = (zz'e\infty) = \frac{z-e}{z'-e} = \frac{z - \dfrac{b}{d-a}}{\dfrac{az+b}{d} - \dfrac{b}{d-a}} =$$

$$\frac{z(d-a)-b}{az(d-a)-ab} \cdot d = \frac{d}{a} = \lambda. \quad (6)$$

A translation is a direct parabolic similitude.

Remark. *For a parabolic homography (similitude or not),* $\lambda = 1$.

Theorem III. *Let* E *and* F *be two fixed points and* λ *a given number, real or imaginary but different from* 0. *If to each point* Z *we associate the point* Z$'$ *such that* (EFZZ$'$) $= \lambda$, *then this correspondence is a Möbius transformation having* E *and* F *for double points.*

1° If E, F are in the finite part of the plane, we have

$$(efzz') = \lambda,$$

$$\frac{z-e}{z-f} = \lambda \frac{z'-e}{z'-f},$$

$$zz'(1-\lambda) + z(\lambda e - f) + z'(\lambda f - e) + ef(1-\lambda) = 0 \quad (7)$$

and since

$$(1-\lambda)^2 ef - (\lambda e - f)(\lambda f - e) = \lambda(e-f)^2 \neq 0,$$

the transformation is a homography. The equation giving the affixes of the double points is

$$(1-\lambda)[z^2-(e+f)z+ef]=0 \tag{8}$$

and is an identity if $\lambda = 1$, in which case (7) is the equation $z = z'$ of the identity transformation. If $\lambda \neq 1$, the roots of (8) are e and f, so that E and F are the double points of the homography.

2° If F is at infinity, we have

$$(e\,\infty\,zz') = (zz'\,e\,\infty) = \lambda,$$

$$z-e=\lambda(z'-e),$$

$$z-\lambda z'+e(\lambda-1)=0,$$

the equation of a similitude with double point E if $\lambda \neq 1$, and the identity transformation if $\lambda = 1$. This result can be gotten from (7) by taking f infinite.

Theorem IV. *If a homography is not a similitude, its equation can be put in the form*

$$(z-e)(z'-f)+(z-f)(z'-e)=(z-z')(m'-l), \tag{9}$$

e, f *being the affixes of the double points* E, F, *distinct or not, and* l, m′ *those of the limit points* L, M′.

Equation (1) or

$$zz'-\frac{a}{c}z+\frac{d}{c}z'-\frac{b}{c}=0$$

can, since

$$e+f=\frac{a-d}{c}, \quad ef=-\frac{b}{c},$$

be written in either one of the two forms

$$zz'-(e+f+\frac{d}{c})z+\frac{d}{c}z'+ef=0, \tag{10}$$

$$zz'-\frac{a}{c}z+(\frac{a}{c}-e-f)z'+ef=0, \tag{11}$$

and is therefore equivalent to the equation obtained by adding (10) and (11), member to member, which gives

$$2zz'-(e+f)(z+z')+2ef=\frac{a+d}{c}(z-z').$$

This equation is nothing but (9) if we take note of (2).

93. Decomposition of a homography. *If a homography is not a similitude, it is a product of similitudes, an inversion, and a symmetry with respect to a line.*

From equation (3) we find

$$z' = m' + \frac{bc - ad}{c^2(z - l)}.$$

If we set

$$z_1 = z - l, \quad z_2 = \frac{1}{z_1},$$

we have

$$z' = m' + \frac{bc - ad}{c^2} z_2.$$

We thus pass from Z to Z_1 by a translation, from Z_1 to Z_2 by an inversion of center O and power 1 followed by a symmetry with respect to Ox (9), and then from Z_2 to Z′ by a similitude.

94. Definitions. The constant (**92**)

$$(\text{EFZZ}') = \lambda$$

is called the *invariant of the homography*. The homography, assumed not to be the identity transformation, is *parabolic* if $\lambda = 1$, *hyperbolic* if λ is real but different from 0 and 1, *elliptic* if λ is imaginary with unit modulus and therefore of the form $e^{i\alpha}$ with α not an integral multiple of π, and *loxodromic* if λ is imaginary with non-unit modulus.

The corresponding cases for the similitude are : translation, homothety, rotation, and product of a rotation and a concentric homothety.

95. Parabolic homography. 1° *The limit points* L, M′ *are distinct and the double point* E *is the midpoint of their join. The line* LM′ *is double for the homography.*

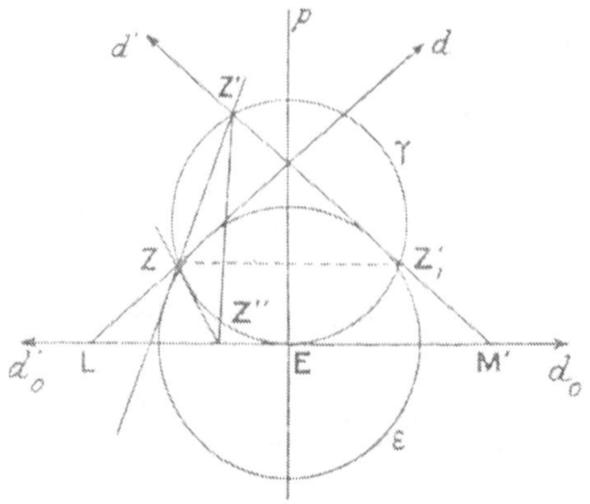

Fig. 47

We have $-d/c \neq a/c$ or $a \neq -d$, for otherwise from $(a - d)^2 + 4bc = 0$ (**92**) we would obtain $a^2 + bc = 0$ or $ad - bc = 0$, which cannot be. Line LM′ is double because line M′E corresponds to line LE (**91, II**) and we know that E is the midpoint of LM′ (**92, I**).

2° *Any two corresponding points* Z, Z′ *are on the circle* γ *passing through* Z *and tangent to line* LM′ *at* E.

The homography is the product of the symmetry with respect to the perpendicular bisector p *of segment* LM′ *and the inversion having center* M′ *and power* $(LM')^2/4$, *or of the inversion* $(L, (LM')^2/4)$ *and the symmetry having axis* p.

We orient the line $LZ = d$ from L toward Z. Since line LM′ is double, point Z′ is (**91**, **III**) on the axis d' which is symmetric to d with respect to p, and is such that we have in magnitude and sign

$$LZ.M'Z' = LE^2.$$

If Z_1' is the symmetric of Z with respect to p, we have

$$M'Z_1'.M'Z' = M'E^2,$$

which establishes the property.

It follows that *every circle* γ *tangent to* LM′ *at* E *is transformed into itself* by the parabolic homography. We say that such a circle is **anallagmatic.**

3° *If the points* Z, Z′ *move on the fixed axes* d, d′, *the line* ZZ′ *envelops the circle* ϵ *having center* E *and tangent to* d, d′.

The equation (9) of the homography is

$$2(z - e)(z' - e) = (z - z')(m' - l) \tag{12}$$

and gives, if we equate the moduli of the two members,

$$|EZ.EZ'| = |ZZ'.LE|.$$

Denoting the radius of γ by R and the distance from E to the line ZZ′ by h we have, for the inscribed triangle EZZ′,

$$|EZ.EZ'| = 2Rh,$$

and hence

$$h = \frac{|ZZ'.LE|}{2R}.$$

But $|ZZ'/2R|$ is the sine of the inscribed angle ZEZ′, which is equal to angle ELZ. Therefore h is equal to the distance from E to d, and ZZ′ is tangent to the circle ϵ.

We also conclude that *in a parabolic homography the quotient*

$$\left|\frac{EZ.EZ'}{ZZ'}\right|$$

is constant and equal to half the distance between the limit points.

4° *The equation of the parabolic homography is*

$$\frac{1}{z-e}-\frac{1}{z'-e}=\frac{2}{l-m'}. \tag{13}$$

The homography is determined by its limit points, or by its double point and a pair of corresponding points.

The double point and an arbitrary point Z *are harmonic conjugates with respect to the homologues* Z′, Z″ *of* Z *in the homography and in the inverse of the homography.*

Equation (13) is obtained from equation (12) by replacing $z-z'$ by $(z-e)-(z'-e)$ and by dividing the two members by $(z-e)(z'-e)(m'-l)$. The equation can be written if we know l and m', for $2e=l+m'$, or if we know e and the affixes q and q' of two corresponding points, for

$$\frac{2}{l-m'}=\frac{1}{q-e}-\frac{1}{q'-e}.$$

In this latter case, the second tangents drawn from Q, Q′ to the circle having center E and tangent to line QQ′ intersect the tangent at E to the circle EQQ′ in the limit points L, M′.

Between the affixes of Z″, Z we have the relation

$$\frac{1}{z''-e}-\frac{1}{z-e}=\frac{2}{l-m'}. \tag{14}$$

By subtracting (13) from (14), member from member, we have

$$\frac{2}{e-z}=\frac{1}{e-z'}+\frac{1}{e-z''}$$

and consequently **(30, II)**

$$(ezz'z'')=-1.$$

96. Hyperbolic homography. 1° *Any two corresponding points* Z, Z′ *and the double points* E, F *lie on a common circle* γ. *Every circle passing through* E *and* F *is anallagmatic.*

In fact, the invariant (EFZZ′) $=\lambda$ is real (28).

2° *The limit points lie on the line* EF, *have the same midpoint as segment* EF, *and are inside or outside the segment* EF *according as* λ *is negative or positive. The line* LM′ *is a double line of the homography.*

We have, in fact,

$$\lambda=(\mathrm{EFL}\,\infty)=\frac{\mathrm{EL}}{\mathrm{FL}}.$$

We shall study in § IV the case where the limit points coincide.

3° *The homography is the product of the symmetry with respect to the perpendicular bisector* p *of segment* LM′ *and the inversion which, having its center at the limit point* M′, *exchanges the double points, or of the similar inversion with center at* L *and the symmetry having axis* p.

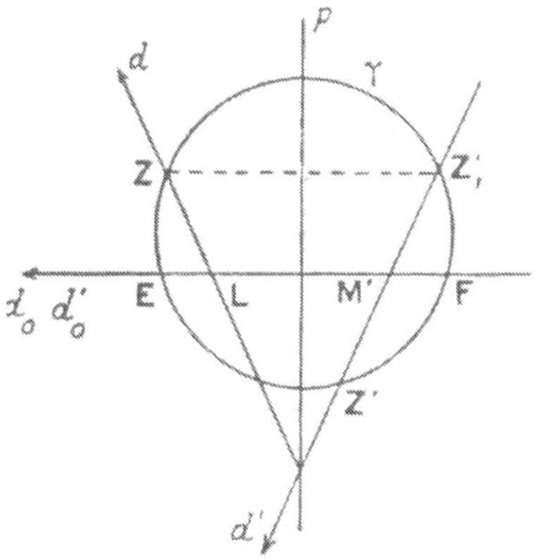

FIG. 48

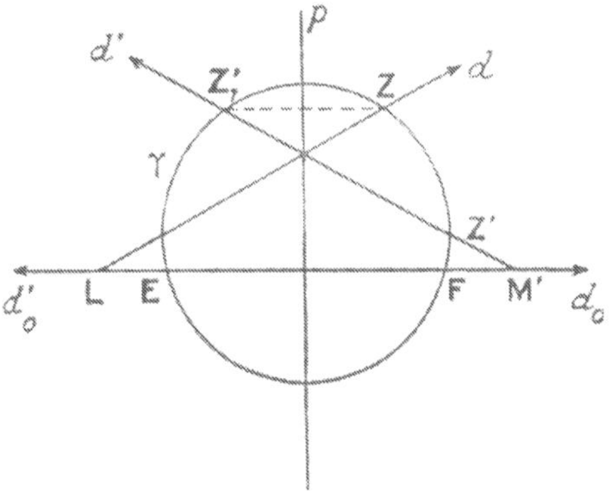

FIG. 49

In the corresponding inversely equal pencils having centers L and M′, the axes d_0, d_0' placed on LM′ are oriented positively from L toward E and from M′ toward E, so that the sense from L toward Z for d fixes the sense of d', which is that from M′ toward Z′. We then reason as in 2° of article **95**.

97. Elliptic homography. We have

$$\lambda = (\mathrm{EFZZ'}) = (\mathrm{EFL}\infty) = (\mathrm{EF}\infty\mathrm{M'}) = e^{i\alpha}. \qquad (15)$$

1° *The limit points* L, M′ *lie on the perpendicular bisector* q *of segment* EF, *and* $(\overline{\mathrm{LF}}, \overline{\mathrm{LE}}) = \alpha$.

2° *Any two corresponding points* Z, Z′ *lie on a circle* γ_1 *of the pencil of circles having* E, F *for limit points. All the circles of this pencil are anallagmatic.*

In fact, from (15) we find

$$\left| \frac{\mathrm{ZE}}{\mathrm{ZF}} \right| = \left| \frac{\mathrm{Z'E}}{\mathrm{Z'F}} \right| .$$

3° *The line* LM′ *is a double line of the homography*, for the axes $d_1 = \mathrm{LE}$ and $d_1' = \mathrm{M'E}$ are homologues.

4° *The homography is the product of the symmetry with respect to the perpendicular bisector* p *of segment* LM′ *and the inversion having center at the limit point* M′ *and power equal to the square of the distance*

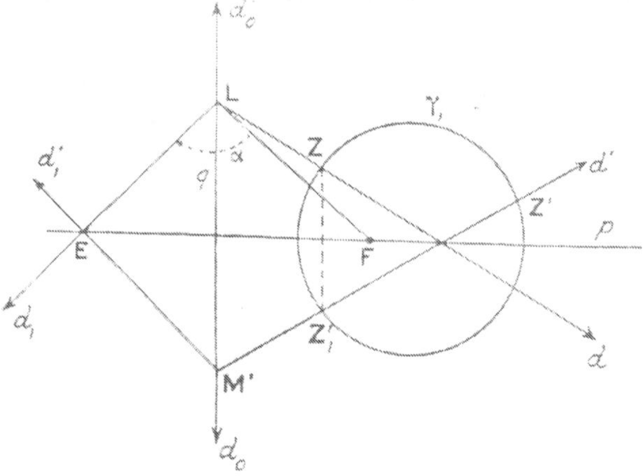

FIG. 50

from the center to a double point, or of the similar inversion having center at L *and the symmetry having axis* p.

In fact, LZ.M′Z′ = $M'Z'_1$.M′Z′ = M′F².

98. Siebeck's theorem.[1] *The double points* E, F *of a direct circular transformation are the foci of the conic enveloped by the line joining two variable corresponding points* Z, Z′ *lying on two corresponding lines* d, d′ *radiating from the limit points* L, M′.

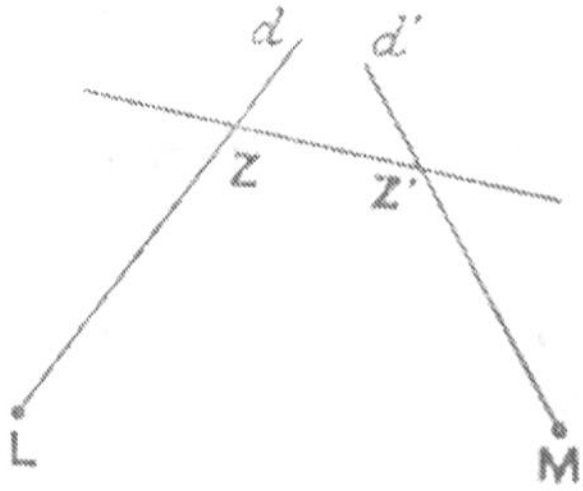

FIG. 51

The transformation determines on *d* and *d′* two projective point ranges which are not perspective if the point *dd′* is not E or F. Line ZZ′ then envelops a conic γ tangent to *d* and *d′*. The center of γ is the midpoint of LM′, for, L and M′ being the limit points of the ranges (Z) and (Z′), the parallels to *d′* and *d* drawn through L and M′ are tangents to γ.

The pencils E(Z), E(Z′) are projective and their double rays are the tangents to γ drawn from E. It follows that E will be a focus

[1] *Archiv der Mathematik und Physik*, vol. 33, 1859, pp. 462-474. Its proof is based on calculus.

if these tangents are isotropic lines, or if the pencils are directly equal, that is, if the angle (EZ, EZ′) is constant to within a multiple of π.

Let Z_1, Z_1' be two other corresponding points on d, d'. From

$$(ZZ_1E\infty) = (Z'Z_1'EM')$$

we obtain, by equating the arguments,

$$(EZ_1, EZ) = (EZ_1', EZ') \pm \pi,$$

$$(EZ_1, EZ') + (EZ', EZ) = (EZ_1', EZ_1) + (EZ_1, EZ') \pm \pi,$$

$$(EZ', EZ) = (EZ_1', EZ_1) \pm \pi,$$

which is what we had to establish.[1]

Corollary. *The conic γ is a circle if the homography is parabolic.* We have already established this (**95**, 3°).

Exercises 84 through 91

84. Steiner ellipses. The conic tangent to the sides of a triangle ABC at their midpoints is an ellipse $\mathscr{E}$ called the *inscribed Steiner ellipse* of the triangle; it has the centroid G of ABC for center. The homothety with center G and coefficient -2 transforms the inscribed Steiner ellipse into the *circumscribed Steiner ellipse.*

1° The tangents to $\mathscr{E}$ form two projective point sets on AB and AC which determine a homography ω_a in the Gauss plane, whose limit points L, M′ are defined by

$$AL : LB = 2, \qquad AM' : M'C = 2;$$

B and C are corresponding points.

2° If a, b, c are the affixes of A, B, C in any rectangular system, the equation of ω_a is

$$3zz' - (a + 2c)z - (a + 2b)z' + ab + bc + ca = 0.$$

[See article **91**, theorem I.]

3° With the notation of exercise 26, show (article **98**) that the affixes of the foci F_1, F_2 of the ellipse $\mathscr{E}$ are the roots of the equation in ϕ

$$3\phi^2 - 2\sigma_1\phi + \sigma_2 = 0.$$

If a, b, c are the roots of

$$z^3 - \sigma_1 z^2 + \sigma_2 z - \sigma_3 = 0,$$

the equation obtained from this by differentiation is none other than the equation in ϕ [2].

---

[1] We have given another proof in *Mathesis*, 1932, p. 268.

[2] Beltrami (*Memorie della Acc. di Bologna*, vol. IX, 1869, pp. 607-657) considered F_1, F_2 as the polar pair of the point at infinity with respect to the triple A, B, C, and F. Morley (*Quarterly Journal*, vol. XXV, 1891, pp. 186-197) identified them with the foci of $\mathscr{E}$. [See *Mathesis*, 1955, p. 81.]

4° By taking the origin at G and Gx on GF_1, show that if θ_a, θ_b, θ_c are the angles that GA, GB, GC make with GF_1, then we have

$$3GF_1^2 = -[GA \cdot GB \cos(\theta_a + \theta_b) + GB \cdot GC \cos(\theta_b + \theta_c) + GC \cdot GA \cos(\theta_c + \theta_a)],$$

$$\frac{\sin(\theta_a + \theta_b)}{GC} + \frac{\sin(\theta_b + \theta_c)}{GA} + \frac{\sin(\theta_c + \theta_a)}{GB} = 0.$$

85. The equation of a parabolic homography can always be put in the form (article 95, 4°)

$$\frac{1}{z} - \frac{1}{z'} = \frac{1}{a}, \qquad a \text{ real}.$$

Deduce (article 30, II) that, if A, B are two given points, any point Z and the harmonic conjugate of B with respect to A and to the symmetric of A with respect to Z correspond in a parabolic homography. Determine the double point and the limit points of the homography.

86. Show that if λ is the invariant of a homography having distinct limit points, L, M′, the equation of the homography is

$$(z - l)(z' - m') = -\frac{\lambda}{(\lambda + 1)^2}(l - m')^2.$$

Deduce that the homography with equation

$$(z - l)(z' - m') = k(l - m')^2, \qquad l \neq m',$$

is parabolic, elliptic, or hyperbolic according as k is equal to, less than, or greater than $-1/4$, and loxodromic if k is not real.

When $l = m'$, $\lambda = -1$. For the origin at L, and if E is a double point, the equation is

$$zz' = e^2.$$

87. Invariant of a direct similitude. The invariant is 1 for a translation, imaginary with unit modulus for a rotation of angle different from $(2k + 1)\pi$, real and different from 1 and 0 for a homothety (-1 for a symmetry), imaginary with non-unit modulus for the remaining case. These similitudes can then be called parabolic, elliptic, hyperbolic, and loxodromic, respectively.

The anallagmatic lines and circles are, respectively, the parallels to the translation vector, the circles centered at the center of rotation, the lines drawn through the center of homothety and, if the homothety is a symmetry, the circles centered at the center of homothety. There are no anallagmatic lines or circles in the last case.

88. In a loxodromic homography, the point of intersection of two corresponding lines describes an equilateral hyperbola which passes through the double points E, F and on which the limit points L, M′ are diametrically opposite one another. [This follows immediately by article 91, III. For a direct calculation, take the origin at the midpoint of LM′ and the Ox axis on LM′. Then we have (exercise 86)

$$(z - l)(z' + l) = 4kl^2$$

for the equation of the homography. A line through L has equation

$$z - l = (\bar{z} - l)e^{it},$$

from which we obtain its transform and, if we set

$$\frac{k}{\bar{k}} = a^2 = e^{2i\alpha},$$

the equation

$$z = \frac{l[1 - 2e^{it} + e^{2i(t-\alpha)}]}{1 - e^{2i(t-\alpha)}}$$

of the locus, which, for

$$t - \alpha = t_1, \qquad e^{it_1} = \frac{1 + iT}{1 - iT},$$

can be written as

$$z = -\frac{li(1 + a)T}{2} + \frac{li(1 - a)}{2T}.$$

See article **63** and exercise 51. The affixes ϕ of the foci are given by

$$\phi^2 = l^2\left(1 - \frac{k}{\bar{k}}\right).]$$

89. 1° The cyclic homography of period 3 (exercise 74)

$$\omega = \begin{pmatrix} A & B & C \\ B & C & A \end{pmatrix}$$

has the isodynamic centers W, W′ of the system A, B, C for double points. [For, in order to have (ABCE) = (BCAE), we must have (article **32**) E = W or W′.]

2° Starting with exercise 72, show that the affixes of W, W′ are the roots of equation (3) of exercise 26.

3° The invariant of ω is $\lambda = e^{\pm 2i\pi/3}$ ($\pm$ according to the order in which we choose W, W′ for writing λ), and ω is elliptic. [Consider the product $(ww'ab)(ww'bc)(ww'ca)$.]

4° The limit points L, M′ are the centers of the equilateral triangles constructed on WW′ as side (the *Beltrami points*). [Consider (WW′L∞) = λ.] Their affixes are

$$l = \frac{3\sigma_3 - (ab^2 + bc^2 + ca^2)}{3\sigma_2 - \sigma_1^2},$$

$$m' = \frac{3\sigma_3 - (ac^2 + ba^2 + cb^2)}{3\sigma_2 - \sigma_1^2}.$$

5° When A, B, C are vertices of a triangle inscribed in a circle (O), the inverses of L, M′ in (O) are called the *Brocard points* Ω, Ω' of the triangle. If (O) is the unit circle, we have (exercises 40 and 41)

$$\frac{1}{\omega} = \frac{3 - \left(\frac{b}{c} + \frac{c}{a} + \frac{a}{b}\right)}{3s_1 - s_2\bar{s}_1},$$

$$\frac{1}{\omega'} = \frac{3 - \left(\frac{c}{b} + \frac{a}{c} + \frac{b}{a}\right)}{3s_1 - s_2\bar{s}_1},$$

$$\frac{1}{\omega} + \frac{1}{\omega'} = \frac{9s_3 - s_1s_2}{3s_1s_3 - s_2^2} = \frac{2}{k}.$$

6° Deduce from 5° (and from exercise 41, 8°) that Ω, Ω', which are symmetric to one another with respect to the line joining O to the Lemoine point K, are on the circle of diameter OK (the *Brocard circle*), and that line LM′ (the *Lemoine line*) is the polar of K in (O). The angle (OK, OΩ) is the *Brocard angle* of triangle ABC.

90. A homography admits an irreducible cyclic set of order n only if its invariant has the form

$$\lambda = e^{2n_1\pi i/n},$$

where n_1 is a positive integer less than n and prime to n, and it suffices to take $n_1 < n/2$. Each point of the plane is then the start of a cyclic set of order n.

If $n > 2$, the homography is elliptic. Each cyclic set is situated on an anallagmatic circle and constitutes a *harmonic* (*Casey*) *polygon*, transformed from a regular polygon by an inversion whose center is a double point of the homography. [See *Mathesis*, 1889, p. 50 and 1932, p. 275.]

91. Point P being an arbitrary point in the plane of triangle ABC, we construct the triangles APC′, BPA′, CPB′ directly similar, respectively, to triangles ABC, BCA, CAB.

1° Triangle A′B′C′ corresponds to triangle ABC in a direct similitude having equation

$$z(p - r) + z'(r - s) + ww' - pr = 0,$$

where w, w', r, s are the affixes of the isodynamic centers W, W′ and of the Beltrami points (exercise 89) R, S of triangle ABC.

2° Discuss the nature of ω as P varies, ABC remaining fixed. [See *Mathesis*, 1937. p. 46.]

IV. MÖBIUS INVOLUTION

99. Equation. *A homography of the complex plane*

$$z' = \frac{az + b}{cz + d} \tag{1}$$

is involutoric, and is called a Möbius involution, if

$$a + d = 0. \tag{2}$$

In fact, it is necessary and sufficient that equation (1) be identical to the inverse homography (23)

$$z = \frac{az' + b}{cz' + d}.$$

If we write the two homographies as

$$czz' - az + dz' - b = 0, \qquad czz' + dz - az' - b = 0$$

we conclude that (2) is the sought condition.

From the form

$$czz' - a(z + z') - b = 0$$

we see that *an equation of the involution is symmetric in z and z'.*

Corollary. *If Z' is the homologue of Z, then Z is the homologue of Z'.* We say that Z and Z' *doubly correspond*, or are *conjugate*, in the involution.

100. Sufficient condition. *In order that a homography be involutoric, it is sufficient that two distinct homologous points doubly correspond.*

In fact, if Z_0, Z_0' are two such homologous points, we have the identities

$$cz_0z_0' - az_0 + dz_0' - b \equiv 0,$$

$$cz_0'z_0 - az_0' + dz_0 - b \equiv 0,$$

which, by subtracting member from member, give

$$(a + d)(z_0' - z_0) \equiv 0$$

or, since $z_0 \neq z_0'$, gives $a + d = 0$.

101. Properties. 1° *A Möbius involution is a hyperbolic homography whose invariant has the value* — 1.

It suffices to set $a + d = 0$ in equation (5) of article **92** to find that $\lambda = -1$. This fact is immediate in the case of an involutoric similitude (89).

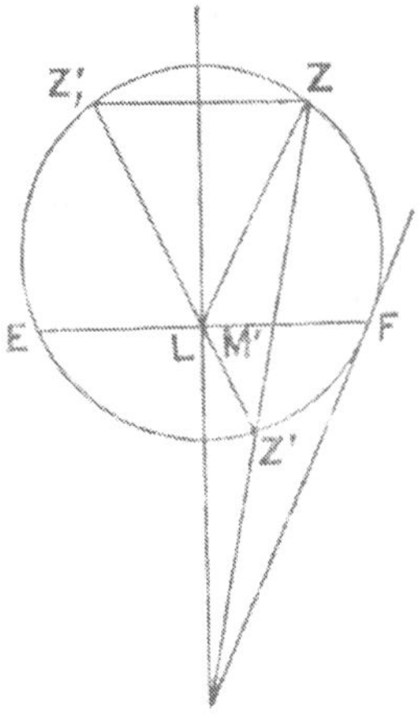

FIG. 52

2° *The double points* E, F *and any two conjugate points* Z, Z' *are vertices of a harmonic quadrangle* (30).

3° *The limit points* L, M' *coincide at the midpoint of segment* EF, the conjugate of the point at infinity of the Gauss plane, and called the *central point* of the involution.

4° *The product of the distances of the central point from two conjugate points is a constant called the power of the involution, and is equal to the square of the distance from the central point to a double point. The line determined by the double points interiorly bisects the angle formed by the lines joining the central point to the two conjugate points* (**30**, **91**).

5° *The line* EF *and its perpendicular at* L *are double lines for the Möbius involution.* The first line carries a hyperbolic involution of points having E, F for double points; the second line carries an elliptic involution of power — LE^2.

6° *The circles passing through* E, F, *as well as the orthogonal trajectories of these circles, are anallagmatic.* In fact, the orthogonal trajectory passing through Z also passes through Z′ (**30**).

7° *A Möbius involution is the product of the inversion of center* L *and power* — LE^2 *and the symmetry with respect to the perpendicular bisector of* EF. *These transformations are permutable* (**96**).

102. Determination of an involution. *An involution is determined if we are given any two pairs of conjugate points; the points of one or of both pairs may coincide.*

Let Z_1, Z_1' and Z_2, Z_2' be the two given pairs of points. If $Z_1 \equiv Z_1'$, $Z_2 \equiv Z_2'$, then these points are the double points of the involution, and the involution is determined inasmuch as the homologue of any point Z is the harmonic conjugate of Z with respect to the double points (**101**). If Z_1 is distinct from Z_1', there exists a unique homography of the complex plane in which Z_1, Z_1', Z_2 have for homologues Z_1', Z_1, Z_2' (**74**), and it is an involution inasmuch as the distinct points Z_1, Z_1' correspond doubly (**100**).

Equation. First form. If z_1, z_1' and z_2, z_2' are the affixes of the given points, the involution has an equation of the form (**99**)

$$azz' + b(z + z') + c = 0$$

with the conditions

$$az_1z_1' + b(z_1 + z_1') + c = 0,$$

$$az_2z_2' + b(z_2 + z_2') + c = 0.$$

The sought equation is then

$$\begin{vmatrix} zz' & z + z' & 1 \\ z_1z_1' & z_1 + z_1' & 1 \\ z_2z_2' & z_2 + z_2' & 1 \end{vmatrix} = 0. \qquad (3)$$

Second form. *If z_1, z_1' and z_2, z_2' are given as roots of the equations*

$$a_1z^2 + 2b_1z + c_1 = 0, \tag{4}$$

$$a_2z^2 + 2b_2z + c_2 = 0, \tag{5}$$

the affixes of an arbitrary pair of conjugate points in the involution are the roots of the equation

$$a_1z^2 + 2b_1z + c_1 + \lambda(a_2z^2 + 2b_2z + c_2) = 0, \tag{6}$$

in which λ is a parameter able to take on all complex values.

The usual equation

$$\alpha zz' + \beta(z + z') + \gamma = 0$$

of an involution says that there exists a constant linear relation between the product and the sum of the affixes of two conjugate points. For Z_1, Z_1' and Z_2, Z_2' we must then have

$$\alpha c_1 - 2\beta b_1 + \gamma a_1 = 0, \tag{7}$$

$$\alpha c_2 - 2\beta b_2 + \gamma a_2 = 0, \tag{8}$$

and for any conjugate pair of points whose affixes are the roots of

$$az^2 + 2bz + c = 0, \tag{9}$$

we will have

$$\alpha c - 2\beta b + \gamma a = 0. \tag{10}$$

From (7), (8), (10) we obtain

$$\begin{vmatrix} a_1 & b_1 & c_1 \\ a_2 & b_2 & c_2 \\ a & b & c \end{vmatrix} = 0$$

and consequently, if μ_1, μ_2 are two arbitrary complex numbers,

$$a = \mu_1a_1 + \mu_2a_2, \quad b = \mu_1b_1 + \mu_2b_2, \quad c = \mu_1c_1 + \mu_2c_2.$$

It suffices to substitute these values into (9) and to set $\mu_2/\mu_1 = \lambda$ in order to obtain (6).

Affixes of the double points. If an involution is given by (3), the affixes of the double points are the roots of

$$\begin{vmatrix} z^2 & 2z & 1 \\ z_1z_1' & z_1 + z_1' & 1 \\ z_2z_2' & z_2 + z_2' & 1 \end{vmatrix} = 0.$$

In the case of equation (6), the λ's for the double points annul the discriminant

$$(\lambda a_2 + a_1)(\lambda c_2 + c_1) - (\lambda b_2 + b_1)^2$$

and the sought affixes are the values of

$$-\frac{\lambda b_2 + b_1}{\lambda a_2 + a_1}$$

for each of the λ's found.

But we can also form an equation of the second degree having these affixes e, f for roots. Let it be

$$rz^2 + 2sz + t = 0. \qquad (11)$$

Since we must have

$$(efz_1z_1') = -1 \quad \text{or} \quad (30, \text{II}) \qquad (e+f)(z_1 + z_1') = 2(ef + z_1z_1'),$$

it is necessary that

$$rc_1 - 2sb_1 + ta_1 = 0 \qquad (12)$$

and, similarly,

$$rc_2 - 2sb_2 + ta_2 = 0. \qquad (13)$$

From (11), (12), (13) we obtain the sought equation

$$\begin{vmatrix} z^2 & -z & 1 \\ c_1 & b_1 & a_1 \\ c_2 & b_2 & a_2 \end{vmatrix} = 0.$$

103. Theorem. *If the four sides of a complete quadrilateral are in the finite part of the plane, the three pairs of opposite vertices are three pairs of conjugate points in a Möbius involution.*

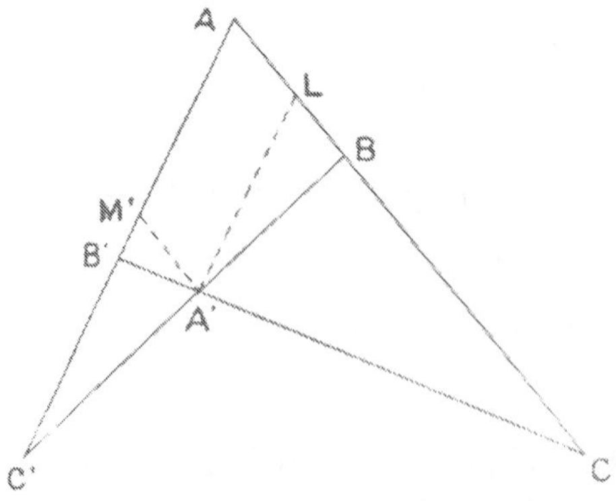

FIG. 53

Let (A, A'), (B, B'), (C, C') be the three pairs of opposite vertices, A, B, C lying on a common side. At least two of these pairs are composed of proper points; let A, A' be one of the pairs. There exists a unique homography ω of the complex plane in which A, B, C have A, C', B' for homologues; A is a double point in this homography. The homography determines on the corresponding lines ABC, AB'C' two projective ranges

$$\text{ABC} \ldots \;\overline{\wedge}\; \text{AC'B'} \ldots$$

which, having point A in common, are perspective from center A'. The limit points L and M' of these ranges, lying on the parallels to AB' and AB drawn through A', are also the limit points of ω. The second double point of ω is thus A' (**92, I**) and we have (**75**) (AA'BC) = (AA'C'B'). Hence we also have (**26**)

$$(AA'BC) = (A'AB'C'),$$

an equation which proves that in the homography of the complex plane which assigns A', B', C' as the homologues of A, B, C, the points A, A' doubly correspond; this homography is then an involution (**100**).

Corollary. *If* C, C' *are at infinity, the involution is a symmetry with respect to the center of the parallelogram* ABA'B' (**89**).

104. Construction of the involution defined by two pairs of points AA', BB'.

It suffices to construct the double points E, F, for then any pair of corresponding points Z, Z' can be constructed by (**30**) (EFZZ') = —1.

Case 1, A, A', B, B' *collinear.* *a*) If the pairs AA', BB' do not separate one another (Fig. 54), E, F are the double points of the hyperbolic involution (AA', BB') determined on the double line AA'BB'. The radical axis *a* of any two circles α, β passing through A, A' and B, B' intersects this line at the central point O. The circle γ of center O and orthogonal to α contains E, F.

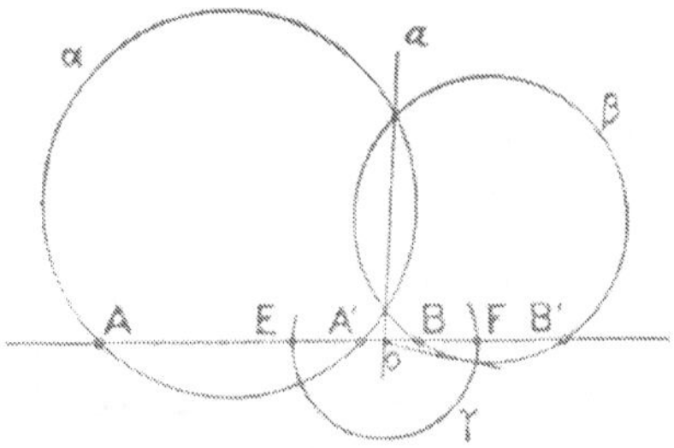

FIG. 54

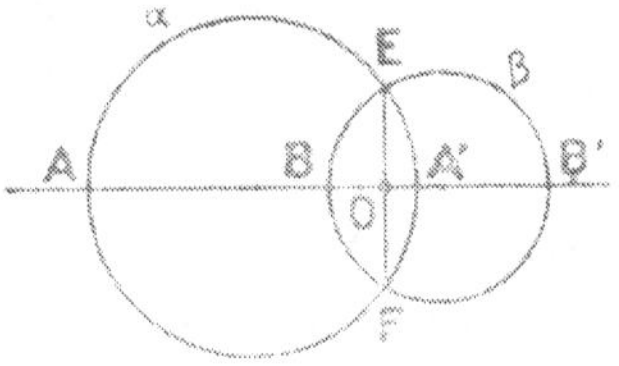

FIG. 55

b) If AA', BB' separate one another, the double line AA'BB' is the perpendicular bisector of the segment EF, and the points E, F are common to the circles α, β having AA', BB' for diameters (Fig. 55).

Case 2, A, A′, B *collinear* (Fig. 56). The line AA′B transforms, by the involution, into the circle γ = A′AB′, which thus contains the central point O. In order to determine O, we note that

$$(BAA'\infty) = (B'A'AO) = (OAA'B'),$$

whence, if P is the second point of γ on the parallel to AA′B drawn through B′, we have

$$P(BAA'\infty) = P(OAA'B')$$

and point O is on the line PB.

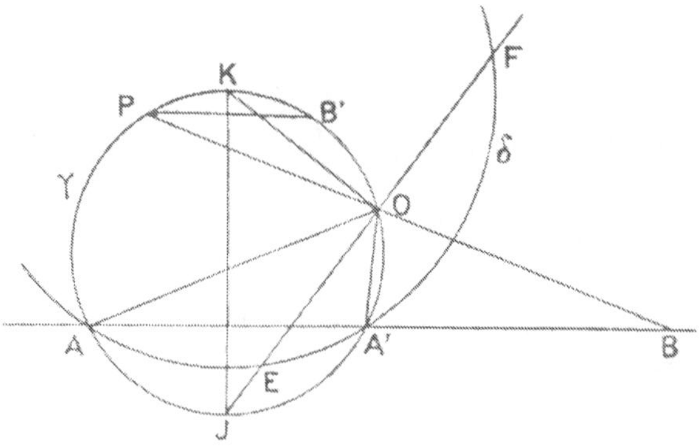

Fig. 56

The problem becomes that of constructing E, F such that (EFAA′) = — 1 and such that O shall be the midpoint of EF (**31**). The interior bisector OJ of angle AOA′ is cut in E, F by the circle δ passing through A and A′ and having its center K on the exterior bisector of this angle.

Case 3, A, A′, B, B′ *are the vertices of a quadrangle.* *a*) If A′ is at infinity, then A is the central point and E, F are such that (EFBB′) = — 1 with A as the midpoint of EF (**31**).

b) If the four points are proper points, let C, C′ be the points (AB, A′B′), (AB′, BA′). Points C, C′ are conjugate points in the involution, for they form the third pair of opposite vertices of the quadrilateral having sides *a* = AB′C′, *b* = BC′A′, *c* = CA′B′, *d* = ABC (**103**). The involution transforms the lines *a*, *b*, *c*, *d* into the circles *a′* = A′BC, *b′* = B′CA, *c′* = C′AB, *d′* = A′B′C′, which then have in common the central point of the involution (**91, II**). The double points E, F are the intersections of the interior bisector *p* of angle BOB′, for example, with the circle β

passing through B, B′ and having its center on the exterior bisector *q* (31).

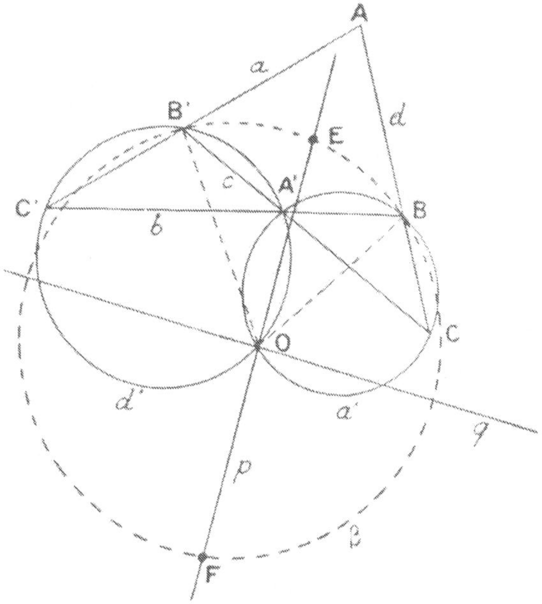

FIG. 57

Corollary. *In a complete quadrilateral* abcd, *the circles circumscribed about triangles* abc, bcd, cda, dab *have a common point* O, *called the* **Miquel point** *of the quadrilateral.*

The angles subtended at O *by the diagonals* AA′, BB′, CC′ *have the same interior bisector* p.

The circles passing through the pairs of opposite vertices (A, A′), (B, B′), (C, C′) *and centered on the exterior bisector* q *have two common points* E, F *on* p, *and we have* (**91, III**)

$$OA.OA' = OB.OB' = OC.OC' = OE^2 = OF^2.$$ [1]

[1] See our study, *Involution de Möbius et point de Miquel: Mathesis*, 1945, vol. LV, pp. 223-230.

Exercises 92 through 96

92. Being given an equilateral triangle ABC, show that the homography having B, C for double points and A for a limit point is cyclic of period 6, and that its cube is a Möbius involution. Construct the cyclic set which belongs to a given point P.

93. Two Möbius involutions with equations

$$a_i zz' + b_i(z + z') + c_i = 0, \qquad i = 1, 2,$$

have in common a pair of conjugate points whose affixes are required. Construct the pair when the involutions are given by their double points.

94. Being given a Möbius involution and a point P, there exists only one pair of points of the involution having P for midpoint. Construct them.

95. Let us refer a Möbius involution to two rectangular axes Ox, Oy whose origin is the central point of the involution and such that the Ox axis contains the double points E, F with affixes a and $-a$.

1° If x, y and x', y' are the coordinates of two conjugate points Z, Z′, then the cartesian equations of the involution are

$$x' = \frac{a^2x}{x^2 + y^2}, \qquad y' = -\frac{a^2y}{x^2 + y^2}.$$

2° Ox and Oy are the only lines which contain an infinite number of pairs of points of the involution. For a line to contain a pair of points of the involution, it is necessary and sufficient that the line intersect the segment EF; construct such a pair of points.

3° The pairs of points of the involution collinear with a given point (p,q) describes the circular cubic having equation

$$(x^2 + y^2)(qx - py) + 2a^2xy - a^2(qx + py) = 0.$$

96. The homography

$$\begin{pmatrix} E & F & A & B & C & \dots \\ E & F & A' & B' & C' & \dots \end{pmatrix}$$

implies the involutions

(EF, AB′, BA′), (EF, AC′, CA′), (EF, BC′, CB′), ...

[Use articles **75** and **100**.]

V. PERMUTABLE HOMOGRAPHIES

105. Sufficient condition. *A sufficient condition for two homographies of the complex plane to be permutable is that they have the same double points.*

We know that for two similitudes the condition is both necessary and sufficient (88).

Parabolic homographies. If E is the double point common to the parabolic homographies with equations (**95**)

$$\frac{1}{z-e}-\frac{1}{z'-e}=a_1, \quad \frac{1}{z'-e}-\frac{1}{z''-e}=a_2 \qquad (1)$$

which permit us to pass from point Z to point Z′, then from Z′ to Z″, the homography associating Z with Z″ has the equation

$$\frac{1}{z-e}-\frac{1}{z''-e}=a_1+a_2 \qquad (2)$$

and is *parabolic with double point* E. If we interchange the roles of the first two homographies, which amounts to interchanging the constants a_1, a_2 in equation (1), the product is again given by (2).

Non-parabolic homographies. Let E, F be the common double points and let λ_1 and λ_2 be the invariants of the two homographies ω_1, ω_2. The equations of these homographies are (**92, III**)

$$(efzz')=\lambda_1, \quad (efz'z'')=\lambda_2$$

and the equation of $\omega_2\omega_1$ is

$$(efzz')\,(efz'z'')=(efzz'')=\lambda_1\lambda_2.$$

The product is thus a homography with double points E, F *and invariant* $\lambda_1\lambda_2$; it is therefore the same as $\omega_1\omega_2$.

106. *The homographies permutable with a parabolic homography* ω_1 *with double point* E *are the parabolic homographies with double point* E.

Any homography ω_2 permutable with ω_1 must have E for double point. Otherwise, let E_2 be the homologue of E in ω_2; it will also be the homologue of E in $\omega_2\omega_1$ (**21**), and can be the homologue of E in $\omega_1\omega_2$ only if E_2 is double for ω_1, which is impossible since E is the only double point of ω_1.

Homography ω_2 cannot have a double point F_2 distinct from E. In fact, the homologue F_{21} of F_2 in ω_1 is distinct from E and F_2, and is also the homologue of F_2 in $\omega_1\omega_2$; it cannot, on the other hand, be the homologue of F_2 in $\omega_2\omega_1$ since it is not double for ω_2.

The sought ω_2's must then have the unique double point E, which is sufficient (**105**).

107. *A non-involutoric homography* ω_1 *with double points* E, F *is permutable with only one involution* I, *namely that which has* E, F *for double points.*

If E is not double for I, its conjugate E′ in I cannot be distinct from F, for E′, the homologue of E in $I\omega_1$, would not be the homologue of E in $\omega_1 I$ as it should be if ω_1 and I are permutable.

If E′ is at F, it is the homologue of E in $I\omega_1$ and in $\omega_1 I$. But then the conjugate of F in I is E, and the double points X, Y of I are such that $(EFXY) = -1$. The homologue X_1 of X in ω_1 is the homologue of X in $\omega_1 I$ but not that of X in $I\omega_1$ because ω_1, being non-involutoric, cannot have its invariant $(EFXX_1)$ equal to -1, and X_1 is not double for I. The hypothesis thus does not give an involution I which is permutable with ω_1.

It is therefore necessary that E be double for I and, by the first part of the argument, that F be the second double point of I. This condition is sufficient (**105**).

The involution I is **the united involution** of the homography.

Corollary. *If* A_1 *and* A_{-1} *are the homologues of a point* A *in the homography* ω_1 *and in its inverse* ω_1^{-1}, *the conjugate of* A *in the united involution* I *is its harmonic conjugate* A″ *with respect to* A_1 *and* A_{-1}.

We have the equation (**92, II**)

$$(EFA_{-1}A) = (EFAA_1),$$

or (**26**)

$$(EFA_{-1}A) = (FEA_1A),$$

proving (**100**) that A is double for a Möbius involution admitting the pairs EF, $A_{-1}A_1$. If A″ is the second double point we then have

$$(EFAA'') = -1, \qquad (A_{-1}A_1AA'') = -1,$$

which establishes the corollary.

Remark. When ω_1 is parabolic with double point E, point A″ is always at E (**95**, 4°).

108. Harmonic involutions. *A necessary and sufficient condition for two distinct involutions to be permutable is that the double points of one be harmonic conjugates with respect to the double points of the other.*

We also say that the *involutions* are *harmonic*.

Let I_1, I_2 be two involutions with double points E_1, F_1 and E_2, F_2, and with equations

$$a_1zz_1 + b_1(z + z_1) + c_1 = 0,$$

$$a_2z_1z_2 + b_2(z_1 + z_2) + c_2 = 0.$$

The homography I_2I_1 has for equation (**78**)

$$(a_1b_2 - a_2b_1)zz_2 + (a_1c_2 - b_1b_2)z + (b_1b_2 - a_2c_1)z_2 + b_1c_2 - b_2c_1 = 0 \quad (3)$$

and is identical to the homography I_1I_2 with equation

$$(a_2b_1 - a_1b_2)zz_2 + (a_2c_1 - b_2b_1)z + (b_2b_1 - a_1c_2)z_2 + b_2c_1 - b_1c_2 = 0 \quad (4)$$

if the coefficients of equations (3) and (4) are proportional. Since I_1 and I_2 are distinct, the matrix

$$\begin{Vmatrix} a_1 & b_1 & c_1 \\ a_2 & b_2 & c_2 \end{Vmatrix}$$

is of rank 2. If b_1 or b_2 is not zero, we do not have simultaneously

$$a_1b_2 - a_2b_1 = 0, \qquad b_1c_2 - b_2c_1 = 0,$$

and the quotient of two corresponding coefficients in (3) and (4) is -1. The coefficients of z or of z_2 give

$$a_1c_2 - 2b_1b_2 + c_1a_2 = 0, \quad (5)$$

which is a necessary and sufficient condition that the double points of I_1, I_2 whose affixes are roots of

$$a_1z^2 + 2b_1z + c_1 = 0, \qquad a_2z^2 + 2b_2z + c_2 = 0$$

form a harmonic quadruple (**102**).

If $b_1 = b_2 = 0$, we have $a_1c_2 - a_2c_1 \neq 0$ and (3) and (4) give $I_2I_1 = I_1I_2$ if $a_1c_2 + a_2c_1 = 0$, which is (5).

Corollary. *If two Möbius involutions with double points* E_1, F_1 *and* E_2, F_2 *are harmonic, their product is a new involution with double points* E_3, F_3, *and each pair of these three involutions is harmonic; the quadrangles* $E_1E_2F_1F_2$, $E_2E_3F_2F_3$, $E_3E_1F_3F_1$ *are harmonic* (**30**).

In fact, (5) says that in (3) the coefficients of z, z_2 are equal, and I_2I_1 is an involution I_3. From the equation

$$I_2I_1 = I_3$$

we obtain

$$I_2I_1I_1 = I_3I_1 \quad \text{or (23)} \quad I_2 = I_3I_1.$$

Since the product I_3I_1 is an involution, the reasoning above permits us to conclude that I_1, I_3 are also harmonic. In the same way we show that I_2, I_3 are harmonic.

109. *A necessary and sufficient condition for two non-involutoric homographies to be permutable is that they have the same double points.*

We know that this theorem is true if one of the homographies is parabolic (**106**). If the given homographies ω_1, ω_2 are not parabolic, let E, F be the double points of ω_1. If E is not double for ω_2, its homologue E_2 in ω_2 is also the homologue of E in $\omega_2\omega_1$, and in order that E_2 be at the same time the homologue of E in $\omega_1\omega_2$, it must be double for ω_1, and hence be at F. Similarly, the homologue F_2 of F in ω_2, not being F, will have to be E. Consequently, E, F will correspond doubly in ω_2, which will thus be involutoric, which is contrary to the hypothesis. It is necessary, then, that E be double for ω_2, and we easily conclude that F must be double for ω_2. These conditions are, moreover, sufficient (**105**).

Corollary. *A non-involutoric homography is permutable with the homographies having the same double points, and, in particular, with its united involution* (**107**) *if it is not parabolic.*

A Möbius involution is permutable with the homographies for which it is the united involution, and with the involutions which have its double points for conjugate points (**108**).

110. *A necessary and sufficient condition for the product of a homography* ω *and an involution* I *to be an involution is that the double points of* ω *be conjugate points in* I. *We say that* I *is* **harmonic** *to* ω.

If the equations of ω and I are

$$\alpha zz' + \beta z + \gamma z' + \delta = 0, \tag{6}$$

$$az'z'' + bz' + bz'' + c = 0, \tag{7}$$

that of $I\omega$ is

$$(\alpha b - \beta a)zz'' + (\alpha c - \beta b)z + (\gamma b - \delta a)z'' + \gamma c - \delta b = 0$$

and represents an involution J if

$$\alpha c - \beta b = \gamma b - \delta a$$

or

$$a\frac{\delta}{\alpha} - b\frac{\beta+\gamma}{\alpha} + c = 0. \tag{8}$$

Since the affixes of the double points E, F of ω are the roots of

$$\alpha z^2 + (\beta + \gamma)z + \delta = 0,$$

and have for product and sum δ/α and $-(\beta+\gamma)/\alpha$, equation (8) says that E, F are conjugate points in I.

Corollaries. 1° *If* ω *is parabolic, its double point is also double for* I.

2° *Every homography* ω *is, in a double infinity of ways, the product* *of two Möbius involutions; these have the double points of ω as a common pair of points.*

In fact, from $I\omega = J$ we obtain $\omega = IJ$, since $I^2 = 1$. We determine an involution I by arbitrarily selecting one of its double points. Since the equation $I\omega = J$ can be written as $I = J\omega^{-1}$ and since ω, ω^{-1} have the same double points, we conclude that these points are also conjugate points in J.

111. Simultaneous invariant of two homographies. As generalizations of articles **108** and **110**, we can consider two homographies ω_1, ω_2 with equations

$$\alpha_1 z z_1 + \beta_1 z + \gamma_1 z_1 + \delta_1 = 0,$$

$$\alpha_2 z_1 z_2 + \beta_2 z_1 + \gamma_2 z_2 + \delta_2 = 0$$

whose product $\omega_2\omega_1$, with equation (**78**)

$$(\alpha_1\gamma_2 - \alpha_2\beta_1)zz_2 + (\alpha_1\delta_2 - \beta_1\beta_2)z + (\gamma_1\gamma_2 - \alpha_2\delta_1)z_2 + \gamma_1\delta_2 - \beta_2\delta_1 = 0,$$

is an involution.

It is necessary and sufficient that we have (**99**) equality of the coefficients of z and z_2, or

$$\alpha_1\delta_2 - \beta_1\beta_2 - \gamma_1\gamma_2 + \alpha_2\delta_1 = 0. \qquad (9)$$

The left member of this equation is a *simultaneous invariant* of ω_1, ω_2.

Corollaries. 1° If ω_1, ω_2 are not similitudes and if they have E_1, F_1, E_2, F_2 for double points and L_1, M_1', L_2, M_2' for limit points (**91**), condition (9) becomes

$$e_1 f_1 + e_2 f_2 = l_1 l_2 + m_1' m_2'. \qquad (10)$$

2° Assuming $\omega_2 \equiv \omega_1$, we have the *condition*

$$\beta_1^2 + \gamma_1^2 - 2\alpha_1\delta_1 = 0 \qquad (11)$$

for the square of ω_1 to be a Möbius involution.

If ω_1 is a similitude, then $\alpha_1 = 0$, (11) gives $\gamma_1 = \pm i\beta_1$, and ω_1 is a rotation of angle $\pm \pi/2$.

If $\alpha_1 \neq 0$, (10) becomes

$$2e_1 f_1 = l_1^2 + m_1'^2$$

and, by taking E_1 for origin, $m_1' = \pm i l_1$; ω_1 is an elliptic homography with invariant $\lambda = e^{\pm i\pi/2}$ (**97**).

112. Transform of a homography. In a homography ω_i, the homologue of a point A or A_k will be designated by A_i or A_{ki}.

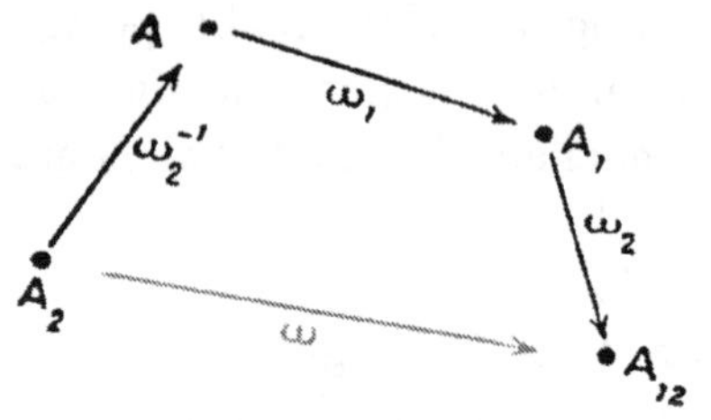

FIG. 58

Let A, A_1 be any two corresponding points of a homography ω_1, and let A_2, A_{12} be the homologues of A, A_1 in a homography ω_2. The correspondence which connects A_2 and A_{12} as A moves in the Gauss plane is the product of the successive homographies ω_2^{-1}, ω_1, ω_2 and is therefore (78) the homography

$$\boxed{\omega = \omega_2\omega_1\omega_2^{-1}.} \tag{12}$$

This homography ω is called the *transform of* ω_1 *by* ω_2.

To any two corresponding points A, A_1 *in the homography* ω_1 *that is transformed, the transforming homography* ω_2 *associates two corresponding points* A_2, A_{12} *in the transformed homography* ω.

In particular, each double point of ω_1 is transformed into a double point of ω. Since, on the other hand, a homography does not alter anharmonic ratios, *the homographies* ω_1 *and* ω *are simultaneously parabolic, hyperbolic, elliptic, or loxodromic with the same invariant, and a Möbius involution is transformed into a Möbius involution.*

The permutability of the homographies ω_1, ω_2 which is expressed by

$$\omega_2\omega_1 = \omega_1\omega_2 \tag{13}$$

is related as follows to the notion of the transform of one homography by another. Multiply the two members of (13) on the right by ω_2^{-1}, then by ω_1^{-1} (21). Since $\omega_2\omega_2^{-1} = \omega_1\omega_1^{-1} = 1$, we have

$$\omega_2\omega_1\omega_2^{-1} = \omega_1, \qquad \omega_2 = \omega_1\omega_2\omega_1^{-1}.$$

By virtue of (12), these equations say that *if* ω_1, ω_2 *are two permutable homographies, each is transformed into itself by the other, whence any two corresponding points in the one are transformed by the other into two points which are also corresponding points in the first homography.*

Exercises 97 through 102

97. The cyclic homographies

$$\begin{pmatrix} A & B & C \\ B & C & A \end{pmatrix}, \qquad \begin{pmatrix} A & B & C \\ C & A & B \end{pmatrix}$$

are inverses of one another. Show that:
1° if

$$(BCAA') = (CABB') = (ABCC') = -1,$$

and if W, W′ are the isodynamic centers of A, B, C (exercises 26, 27, 89), the united involution is [article **107**, corollary]

$$(WW, W'W', AA', BB', CC');$$

2° the foci F_1, F_2 of the inscribed Steiner ellipse (exercise 84) constitute a pair of points of this involution and are therefore on a circle which passes through W, W′.

98. The involutions of exercise 96 are harmonic to the homography.

99. An involution is harmonic to a homography ω (article **110**) if it transforms ω into ω^{-1}.

100. Every homography ω can be transformed into a similitude by a properly chosen homography ω_1. It suffices to take ω_1 such that one of its limit points coincides with a double point of ω.

Obtain, with the aid of exercise 87, the anallagmatic lines or circles of a non-similitude homography.

101. 1° The product of two Möbius involutions

$$J_1 = (E_1E_1, F_1F_1), \qquad J_2 = (E_2E_2, F_2F_2)$$

is a homography ω which is: parabolic if $E_1 = E_2$; elliptic if the pairs E_1, F_1 and E_2, F_2 are collinear or concyclic and separate one another non-harmonically; hyperbolic with a positive invariant if the pairs E_1, F_1 and E_2, F_2 are collinear or concyclic and do not separate one another; hyperbolic with a negative invariant if $(E_1F_1E_2F_2)$ is of the form $e^{i\alpha}$ (α real), and involutoric if

$$(E_1F_1E_2F_2) = -1;$$

loxodromic in all other cases.

2° If $(E_1F_1E_2F_2) = k$, the invariant of ω is

$$\left[\frac{1 - \sqrt{k}}{1 + \sqrt{k}}\right]^2.$$

[See *Mathesis*, 1947, p. 65.]

102. Three Möbius involutions which are harmonic in pairs and none of which is a symmetry have distinct central points A_1, A_2, A_3 and are determined by the arbitrary choice of these points.

If P_1, P_2, P_3 are the conjugates of an arbitrary point P, the triangles $P_1A_2A_3$, $A_1P_2A_3$, $A_1A_2P_3$ are directly similar, the lines A_1P_1, A_2P_2, A_3P_3 are concurrent at the isogonal conjugate of P in the proper or degenerate triangle $A_1A_2A_3$, and we have three equations like

$$|A_1P \cdot A_1P_1| = |A_1A_2 \cdot A_1A_3|.$$

[See *Mathesis*, 1947, p. 67.]

VI. ANTIGRAPHY

113. Definition. An antigraphy is a transformation of the Gauss plane which has an equation of the form

$$\alpha\bar{z}z' + \beta\bar{z} + \gamma z' + \delta = 0, \quad \alpha\delta - \beta\gamma \neq 0 \tag{1}$$

in which α, β, γ, δ are real or imaginary constants, or of the form

$$z' = \frac{a\bar{z} + b}{c\bar{z} + d}, \qquad ad - bc \neq 0. \tag{2}$$

The inequalities are justified as in article **73**.

114. Properties. 1° *An antigraphy is the product of a symmetry with respect to the* Ox *axis and a homography.*

Equation (2), for example, follows from the elimination of z'' from the equations

$$z'' = \bar{z}, \qquad z' = \frac{az'' + b}{cz'' + d} \tag{3}$$

of which the first is the equation of the considered symmetry.

2° *An antigraphy is determined if to three distinct and arbitrarily chosen points* Z_1, Z_2, Z_3, *we associate three distinct and arbitrarily chosen points* Z_1', Z_2', Z_3'.

The proof is accomplished as in article **74**.

3° *The anharmonic ratio of any four points* Z_1, Z_2, Z_3, Z_4 *is equal to the* **conjugate** *of the anharmonic ratio of their homologues* Z_1', Z_2', Z_3', Z_4' *in an antigraphy.*

The first of equations (3) gives

$$(\bar{z}_1\bar{z}_2\bar{z}_3\bar{z}_4) = \overline{(z_1z_2z_3z_4)} = (z_1''z_2''z_3''z_4'')$$

and the second (**75**)

$$(z_1''z_2''z_3''z_4'') = (z_1'z_2'z_3'z_4'),$$

whence

$$\overline{(z_1z_2z_3z_4)} = (z_1'z_2'z_3'z_4') \quad \text{or} \quad (z_1z_2z_3z_4) = \overline{(z_1'z_2'z_3'z_4')}.$$

Corollary. *An antigraphy leaves* **real** *anharmonic ratios invariant.*

4° *The product of a homography and an antigraphy is an antigraphy. The product of two antigraphies is a homography.*

Thus, the product of the homography and the antigraphy having equations

$$\alpha_1 z z_1 + \beta_1 z + \gamma_1 z_1 + \delta_1 = 0, \qquad \alpha_2 \bar{z}' z_1 + \beta_2 z_1 + \gamma_2 \bar{z}' + \delta_2 = 0$$

has for equation (78)

$$(\alpha_1\gamma_2 - \alpha_2\beta_1) z\bar{z}' + (\alpha_1\delta_2 - \beta_1\beta_2) z + (\gamma_1\gamma_2 - \alpha_2\delta_1)\bar{z}' + \gamma_1\delta_2 - \beta_2\delta_1 = 0$$

and is an antigraphy.

A. *ANTISIMILITUDE*

115. Equation. The antigraphy with equation (1) or (2) is called an antisimilitude if $a = c = 0$, and it then has an equation of the form

$$z' = p\bar{z} + q.$$

It is the product of the symmetry with respect to Ox and a similitude (**114**, **81**).

116. Properties. I. *Any antisimilitude has the point at infinity for a double point. It transforms any straight line into a straight line, any circle into a circle, and any triangle into an inversely similar triangle for which the ratio of similitude of the second to the first is* $|\,p\,|$.

As z approaches ∞, so does z'.

The straight line or the circle having equation

$$z = \frac{at + b}{ct + d}, \qquad t \text{ a real parameter} \tag{4}$$

is transformed into the curve having the equation

$$z' = \frac{(p\bar{a} + q\bar{c})t + p\bar{b} + q\bar{d}}{\bar{c}t + \bar{d}}$$

which represents (**41**) a straight line or a circle according as (4) represents a straight line or a circle.

If $Z_1Z_2Z_3$, $Z_1'Z_2'Z_3'$ are two corresponding triangles, we have (**114**, 3°)

$$(Z_1Z_2Z_3\,\infty) = \overline{(Z_1'Z_2'Z_3'\,\infty)}$$

and, by expressing that the moduli are equal while the arguments are opposite,

$$\left|\frac{Z_1Z_3}{Z_2Z_3}\right| = \left|\frac{Z_1'Z_3'}{Z_2'Z_3'}\right|, \qquad (\overline{Z_3Z_2}, \overline{Z_3Z_1}) = -(\overline{Z_3'Z_2'}, \overline{Z_3'Z_1'}).$$

The triangles are thus inversely similar and the ratio of similitude $|Z_1'Z_2'/Z_1Z_2|$ is $|p|$ (82). This is why an antisimilitude is also called an *inverse similitude.*

II. *An antisimilitude is determined by two pairs,* Z_1, Z_1' *and* Z_2, Z_2', *of corresponding points, and has for equation*

$$\begin{vmatrix} z' & \bar{z} & 1 \\ z_1' & \bar{z}_1 & 1 \\ z_2' & \bar{z}_2 & 1 \end{vmatrix} = 0.$$

A necessary and sufficient condition for two triangles $Z_1Z_2Z_3$, $Z_1'Z_2'Z_3'$ *to be inversely similar is*

$$\begin{vmatrix} z_1 & \bar{z}_1' & 1 \\ z_2 & \bar{z}_2' & 1 \\ z_3 & \bar{z}_3' & 1 \end{vmatrix} = 0.$$

III. *An antisimilitude is an inversely conformal transformation; it preserves the magnitude of each angle but reverses its sense.*

This follows from property **I.**

117. Symmetry. 1° *The symmetry with respect to the line* d *having equation* (**37**)

$$az + \bar{a}\bar{z} + b = 0, \qquad b \text{ real}$$

is the antisimilitude with ratio 1 *and with equation*

$$az' + \bar{a}\bar{z} + b = 0 \tag{5}$$

and whose double points are all the points of the line d.

The line d contains the point S with affix $-b/2a$ and is parallel to the vector representing the number $i\bar{a}$ (**37**); it therefore also contains the point T with affix $-b/2a + i\bar{a}$. In order that two points Z′, Z be symmetric with respect to d, it is necessary and sufficient that triangles Z′ST, ZST be inversely similar, or that (**116**)

$$\begin{vmatrix} z' & \bar{z} & 1 \\ -\dfrac{b}{2a} & -\dfrac{b}{2\bar{a}} & 1 \\ -\dfrac{b}{2a} + i\bar{a} & -\dfrac{b}{2\bar{a}} - ia & 1 \end{vmatrix} = 0,$$

an equation which is none other than (5).

2° *The antisimilitude with equation*

$$z' = p\bar{z} + q \tag{6}$$

is a symmetry with respect to a line if $|\,p\,| = 1$ *and* $p\bar{q} + q = 0$.

It is necessary and sufficient that the equation

$$z - p\bar{z} - q = 0 \quad \text{or} \quad \lambda z - \lambda p\bar{z} - \lambda q = 0$$

represent a real straight line, and therefore, if r is a real number, that we have (**37**)

$$\bar{\lambda} = -\lambda p, \quad \lambda q = r.$$

The first equation says that $|\,p\,| = 1$. If $q = 0$, the second equation is realized; if $q \neq 0$, elimination of λ gives $p\bar{q} + q = 0$.

3° *Symmetry with respect to a line is the only involutoric antisimilitude.*

The antisimilitude with equation (6) is an involution if its square, with equation

$$z'' = p\bar{z}' + q = p(\bar{p}z + \bar{q}) + q \quad \text{or} \quad z'' = p\bar{p}z + p\bar{q} + q, \tag{7}$$

is the identity transformation (**23**), that is, if

$$p\bar{p} = 1, \qquad p\bar{q} + q = 0.$$

It is then (see 2°) a symmetry with respect to a line.

4° *An antisimilitude is involutoric when there exists a pair of distinct corresponding points Z, Z′ which correspond doubly.*

Such a pair of points gives the relations

$$z' - p\bar{z} - q = 0, \qquad z - p\bar{z}' - q = 0, \qquad z \neq z'$$

or

$$z' - p\bar{z} - q = 0, \qquad \bar{p}z' - \bar{z} + \bar{q} = 0, \qquad z \neq z'.$$

If the determinant $-1 + p\bar{p}$ of the coefficients of the unknowns is not zero, we find

$$z' = \frac{q + p\bar{q}}{1 - p\bar{p}}, \qquad \bar{z} = \frac{\bar{q} + \bar{p}q}{1 - p\bar{p}}$$

and, contrary to hypothesis, we would have $z' = z$. It is necessary, then, that $p\bar{p} = 1$, and that the condition of compatibility of the equations, $q + p\bar{q} = 0$, hold, whence the property.

118. Double points. *In addition to the point at infinity, an antisimilitude possesses*

1° *a proper double point if the ratio of similitude is different from* 1;

2° *a line of double points if it is a symmetry with respect to a line*;

3° *no proper double point in the case of an inverse equality which is not a symmetry.*

The affixes of the possible double points of the antisimilitude with equation

$$z' = p\bar{z} + q$$

are the roots of the equation

$$z - p\bar{z} - q = 0, \tag{8}$$

and hence also those of the conjugate equation

$$-\bar{p}z + \bar{z} - \bar{q} = 0.$$

By considering these two linear equations in z, $\bar{z}$, we find, if $1 - p\bar{p} \neq 0$,

$$z = \frac{q + p\bar{q}}{1 - p\bar{p}}, \qquad \bar{z} = \frac{\bar{q} + \bar{p}q}{1 - p\bar{p}},$$

whence 1°.

When $p\bar{p} = 1$, or $|p| = 1$, the equations are compatible if

$$p\bar{q} + q = 0.$$

The antisimilitude is the symmetry with respect to the line having equation (8) (**117**, 2° and 1°), whence 2°.

If the equations are incompatible, we have 3°.

119. Construction of the double point E. Suppose the inverse similitude is given by the two pairs of corresponding points Z_1, Z_1' and Z_2, Z_2' such that $|Z_1'Z_2'/Z_1Z_2| \neq 1$.

First method. Since we must have

$$\left|\frac{EZ_1'}{EZ_1}\right| = \left|\frac{EZ_2'}{EZ_2}\right| = \left|\frac{Z_1'Z_2'}{Z_1Z_2}\right|, \tag{9}$$

E is common to the circles γ_1, γ_2 which are the loci of points the ratio of whose distances from Z_1, Z_1' or Z_2, Z_2' is $|Z_1'Z_2'/Z_1Z_2|$. But the center E_d of the direct similitude determined by the same pairs Z_1, Z_1' and Z_2, Z_2' also satisfies equations (9) (**83**) and is common to γ_1, γ_2. Since we can recognize E_d with the assistance of the circle containing the points Z_1, Z_1', $(Z_1Z_2, Z_1'Z_2')$ (**83**), point E is determined by elimination.

Second method. Point E is also a double point for the square of the antisimilitude, the equation of which is (7). This represents a homothety since $p\bar{p}$ is real and different from unity by hypothesis. The lines joining two pairs of corresponding points in this homothety then intersect in E.

Therefore, *if we construct the triangles* $Z_1'Z_2'Z_1''$, $Z_1'Z_2'Z_2''$ *respectively inversely similar to the triangles* $Z_1Z_2Z_1'$, $Z_1Z_2Z_2'$, *the lines* Z_1Z_1'', Z_2Z_2'' *intersect in* E.

Exercises 103 and 104

103. In the application of article **90**, triangles ABC, A'B'C' are always directly similar if ABC is equilateral; if ABC is arbitrary, then the midpoints of its sides form the only triangle A'B'C' directly similar to ABC.

Triangles ABC and A'B'C' are inversely similar only if BA' = A'C and if the angle at A', when conventionally oriented, is twice the Brocard angle of ABC. [Take circle (ABC) for unit circle and see exercise 89.]

104. Let A', B', C' be the symmetrics of the vertices A, B, C of a triangle with respect to the sides BC, CA, AB, and let A_1, B_1, C_1 be the midpoints of B'C', C'A', A'B'. Show that triangles AC_1B_1, C_1BA_1, B_1A_1C are inversely similar to ABC.

From this it can be shown that the construction of triangle ABC, knowing A', B', C' (the *problem of three images*) depends upon a 7th degree equation. [See *Mathesis*, 1935, p. 154.]

B. *NON-ANTISIMILITUDE ANTIGRAPHY*

120. Circular transformation. *An antigraphy is an inversely conformal transformation.*

In fact, it is the product of the symmetry with respect to Ox and a homography (**114**); these two transformations are circular (**116**, **76**) and the first is inversely conformal (**116**) while the second is directly conformal (**77**).

121. Limit points. The antigraphy with equation

$$z' = \frac{a\bar{z} + b}{c\bar{z} + d}, \qquad ad - bc \neq 0, \qquad c \neq 0 \tag{1}$$

possesses two limit points L, M' (**80**) with affixes

$$l = -\frac{\bar{d}}{\bar{c}}, \qquad m' = \frac{a}{c} \tag{2}$$

and its equation (1) or

$$cz'\bar{z} + dz' - a\bar{z} - b = 0 \tag{3}$$

can, by taking note of (2), be written as

$$(\bar{z} - \bar{l})(z' - m') = \frac{bc - ad}{c^2}. \qquad (4)$$

We can establish the following properties as in article **91**.

A necessary and sufficient condition for a straight line or a circle (C) *of the* w *plane to transform into a straight line* (C') *of the* w' *plane is that* (C) *contain the limit point* L *of* w.

The pencil of lines with vertex L *is the only pencil which transforms into a pencil of lines, and this pencil has its vertex at* M'.

If r, θ *are the modulus and the argument of* $(bc - ad)/c^2$, *the product*

$$| LZ.M'Z' |$$

has the constant value r. *The pencils of homologous lines having their vertices at* L, M' *are* **directly** *equal and the angle*

$$(\overline{LZ}, \overline{M'Z'})$$

has the constant value θ.

Remark. *The angle of two axes which intersect in the limit point* L *is conserved in magnitude and sense* (see article **120**).

122. Inversion. *An involutoric antigraphy which is not an axial symmetry* (**117**) *is an inversion whose power may be positive or negative.*

According as the power is positive or negative, the circle of inversion is the locus of the double points of the inversion or the inversion has no double points.

The antigraphy $\bar{\omega}$ with equation (1) is an involution if (**23**) its square $\bar{\omega}^2$ is the identity homography. The equation of $\bar{\omega}^2$, which results from the elimination of z' from (1) and

$$z'' = \frac{a\bar{z}' + b}{c\bar{z}' + d},$$

is

$$(c\bar{a} + d\bar{c})zz'' - (a\bar{a} + b\bar{c})z + (c\bar{b} + d\bar{d})z'' - (a\bar{b} + b\bar{d}) = 0. \quad (5)$$

We have $\bar{\omega}^2 = 1$ if

$$c\bar{a} + d\bar{c} = 0, \qquad (6)$$

$$a\bar{a} + b\bar{c} = c\bar{b} + d\bar{d}, \qquad (7)$$

$$a\bar{b} + b\bar{d} = 0. \qquad (8)$$

Relation (6) implies that

$$c\bar{a} = -d\bar{c}, \qquad \bar{c}a = -\bar{d}c, \qquad c\bar{a}\bar{c}a = d\bar{c}\bar{d}c$$

and consequently, since $c \neq 0$, $a\bar{a} = d\bar{d}$ and relation (7) becomes

$$b\bar{c} = c\bar{b} \quad \text{or} \quad \frac{b}{c} = \frac{\bar{b}}{\bar{c}}. \tag{9}$$

As for equation (8), it can, because of (9), be written as

$$\frac{a}{c}\cdot\frac{\bar{b}}{\bar{c}} + \frac{b}{c}\cdot\frac{d}{\bar{c}} = \frac{b}{c}\left(\frac{a}{c} + \frac{d}{\bar{c}}\right) = 0$$

and is true if (6) holds. Therefore, when $c \neq 0$, we have $\bar{\omega}^2 = 1$ if

$$\frac{d}{c} = -\frac{\bar{a}}{\bar{c}}, \qquad \frac{b}{c} \text{ real.} \tag{10}$$

Equation (3) of $\bar{\omega}$ becomes, by taking note of (10),

$$z'\bar{z} - \frac{\bar{a}}{\bar{c}}z' - \frac{a}{c}\bar{z} - \frac{b}{c} = 0$$

or

$$\left(z' - \frac{a}{c}\right)\left(\bar{z} - \frac{\bar{a}}{\bar{c}}\right) = \frac{a\bar{a}}{c\bar{c}} + \frac{b}{c}.$$

We know (**19**) that this equation represents the inversion whose center has a/c for affix and whose power is $a\bar{a}/c\bar{c} + b/c$.

Corollaries. 1° *The equation*

$$z' = \frac{a\bar{z} + b}{c\bar{z} + d}$$

represents an inversion if

$$c \neq 0, \qquad a\bar{c} + c\bar{d} = 0, \qquad \frac{b}{c} \text{ real,}$$

and an axial symmetry if (**117**)

$$c = 0, \qquad |a| = |d|, \qquad a\bar{b} + b\bar{d} = 0.$$

2° *An antigraphy $\bar{\omega}$ is an inversion or an axial symmetry when there exist two pairs of distinct corresponding points such that the points of each pair correspond doubly (involutoric pairs), or when there exists one such pair and a double point.*

If there exist two involutoric pairs Z_1, Z_1' and Z_2, Z_2', we have (**114, 26**)

$$(Z_1Z_1'Z_2Z_2') = (\overline{Z_1'Z_1Z_2'Z_2}) = (\overline{Z_1Z_1'Z_2Z_2'})$$

and the anharmonic ratio is consequently real. The points Z_1, Z_1', Z_2, Z_2' are therefore on a line or on a circle (**28**). In the first case, if Z_1Z_1', Z_2Z_2' have the same midpoint E, we have a symmetry with axis perpendicular to line Z_1Z_2 at E; if Z_1Z_1', Z_2Z_2' do not have a common midpoint, there exists a point O on the line such that

$$OZ_1 . OZ_1' = OZ_2 . OZ_2',$$

and which is therefore the center of an inversion in which Z_1, Z_1' as well as Z_2, Z_2' are homologues, and this is the only antigraphy enjoying this property, for we know four pairs (Z_1, Z_1'), (Z_1', Z_1), (Z_2, Z_2'), (Z_2', Z_2) of corresponding points.[1] If Z_1, Z_1', Z_2, Z_2' are concyclic, then, according as the lines Z_1Z_1', Z_2Z_2' are not or are parallel, $\bar{\omega}$ is an inversion with center $O \equiv (Z_1Z_1', Z_2Z_2')$ or the symmetry whose axis is the diameter perpendicular to these lines.

Suppose, now, that there exists an involutoric pair Z_1Z_1' and a double point E, elements which determine an $\bar{\omega}$. If Z_1' is at infinity, $\bar{\omega}$ is the inversion with center Z_1 and power Z_1E^2; if E is at infinity, $\bar{\omega}$ is the symmetry with respect to the perpendicular bisector of segment Z_1Z_1'. When Z_1, Z_1', E are collinear and in the finite part of the plane, then, according as E is or is not the midpoint of Z_1Z_1', $\bar{\omega}$ is the symmetry with respect to the perpendicular to Z_1Z_1' at E or the inversion in the circle having for diameter the segment EF, where $(EFZ_1Z_1') = -1$. When Z_1, Z_1', E are not collinear, $\bar{\omega}$ is, according as E is or is not on the perpendicular bisector of Z_1Z_1', the symmetry with respect to this line or the inversion having for center O the point of intersection of line Z_1Z_1' with the tangent at E to the circle EZ_1Z_1' and for power OE^2.

3° *A non-involutoric antigraphy cannot have both an involutoric pair and a double point* (corollary 2°).

It can have one and only one involutoric pair.

To obtain such an antigraphy, it is sufficient to determine it by three pairs (Z_1, Z_1'), (Z_1', Z_1), (Z_2, Z_2') of corresponding points for which Z_1, Z_1', Z_2, Z_2' are neither collinear nor concyclic.

4° *A non-involutoric antisimilitude cannot possess an involutoric pair* (**117**, 4°), for it has a double point at infinity.

123. Non-involutoric antigraphies. A non-involutoric antigraphy $\bar{\omega}$ is called *hyperbolic*, *parabolic*, or *elliptic* according as it

[1] If Z_1' is at infinity, Z_1 is the center of the inversion having power $Z_1Z_2.Z_1Z_2'$.

possesses two double points E, F, a single double point E = F, or an involutoric pair P, Q.

Theorem. *$\bar{\omega}$ is hyperbolic if the homography $\bar{\omega}^2$ is hyperbolic but not a Möbius involution; parabolic if $\bar{\omega}^2$ is parabolic; elliptic if $\bar{\omega}^2$ is elliptic or a Möbius involution.*

$\bar{\omega}^2$ is never loxodromic.

The homography $\bar{\omega}^2$, not being the identity since $\bar{\omega}$ is not involutoric, possesses two distinct or coincident double points (**82**, **83**, **92**) X and Y.

A double point of $\bar{\omega}$ is necessarily double for $\bar{\omega}^2$. But if $\bar{\omega}$ possesses an involutoric pair PQ, the homologues of P, Q in $\bar{\omega}^2$ are P, Q. Therefore the points X, Y are double for $\bar{\omega}$ or else constitute, when they are distinct, the involutoric pair of $\bar{\omega}$. Moreover we know (**122**, corollary 3°) that the existence of such a pair excludes the existence of a double point.

The points X, Y are the images of the roots of the equation

$$(c\bar{a} + d\bar{c})z^2 - (a\bar{a} + b\bar{c} - c\bar{b} - d\bar{d})z - (a\bar{b} + b\bar{d}) = 0 \qquad (11)$$

obtained by setting $z'' = z$ in (5). The discriminant of (11),

$$\Delta = (a\bar{a} + b\bar{c} - c\bar{b} - d\bar{d})^2 + 4(c\bar{a} + d\bar{c})(a\bar{b} + b\bar{d}), \qquad (12)$$

is easily written as

$$\Delta = (a\bar{a} + b\bar{c} + c\bar{b} + d\bar{d})^2 - 4(ad - bc)(\bar{a}\bar{d} - \bar{b}\bar{c}), \qquad (13)$$

and consequently is real for any $\bar{\omega}$.

The invariant (**94**, **92**) of $\bar{\omega}^2$ being

$$\lambda = \frac{a\bar{a} + b\bar{c} + c\bar{b} + d\bar{d} + \sqrt{\Delta}}{a\bar{a} + b\bar{c} + c\bar{b} + d\bar{d} - \sqrt{\Delta}} \qquad (14)$$

is the quotient of two real or two conjugate imaginary numbers. It is therefore a real or an imaginary number with unit modulus. Therefore $\bar{\omega}^2$ is never loxodromic.

1° $\bar{\omega}^2$ is a direct similitude if, considering its equation (5), we have (**81**)

$$c\bar{a} + d\bar{c} = 0.$$

This relation implies (**122**)

$$a\bar{a} = d\bar{d} \qquad (15)$$

and expression (12) becomes

$$\Delta = (b\bar{c} - c\bar{b})^2.$$

We do not have

$$b\bar{c} = c\bar{b},$$

for this equation, along with (15), gives (7), which, with (6), makes $\bar{\omega}$ an inversion (122). Since $b\bar{c} - c\bar{b}$ is thus a non-zero pure imaginary number, we have $\Delta < 0$. From (14) we observe that λ is an imaginary $e^{i\theta}$ or has the value -1; $\bar{\omega}^2$ is consequently a rotation or a symmetry with center X, that is, an elliptic homography or an involution. As for $\bar{\omega}$, it necessarily has X and the point at infinity Y for involutoric pairs, for, not being an antisimilitude, it is unable to have the double point Y; its limit points are coincident at X.

2° Suppose, now, that $\bar{\omega}^2$ possesses two limit points L_2, M'_2. The limit points L, M′ of $\bar{\omega}$ are distinct. Since L_2, M'_2 are, respectively, the homologue of L in $\bar{\omega}^{-1}$ and the homologue of M′ in $\bar{\omega}$, we have

$$\bar{\omega} = \begin{pmatrix} L & \infty & M' & L_2 \\ \infty & M' & M'_2 & L \end{pmatrix}, \qquad \bar{\omega}^2 = \begin{pmatrix} X & Y & L & L_2 & \infty \\ X & Y & M' & \infty & M'_2 \end{pmatrix}.$$

Therefore $\bar{\omega}$ gives (114, 3°)

$$(L_2M'L\infty) = \overline{(LM'_2\infty M')}$$

or (26)

$$(L_2M'L\infty) = \overline{(M'_2LM'\infty)}.$$

This equation shows that L and M′ correspond doubly in an antigraphy having L_2, M'_2 for corresponding points. It follows that (117, 4° and 3°) L_2, M'_2 *are symmetric with respect to the perpendicular bisector* $\mathcal{M}$ *of segment* LM′.

If $\bar{\omega}^2$ is parabolic, its unique double point X = Y, the midpoint of segment $L_2M'_2$ (95), is the only double point E = F of $\bar{\omega}$, which is therefore parabolic.

In the remaining cases, in order to know if X, Y are double or are associated involutorically in $\bar{\omega}$, it is sufficient to find out if we have

$$(XYL\infty) = \overline{(XY\infty M')} \tag{16}$$

or

$$(XYL\infty) = \overline{(YX\infty M')}. \tag{17}$$

If $\bar{\omega}^2$ is hyperbolic but is not a Möbius involution, X, Y are on the line $L_2M'_2$ and symmetric with respect to $\mathcal{M}$ (96). This symmetry gives us

$$(XYL\infty) = \overline{(YXM'\infty)} = \overline{(XY\infty M')},$$

that is to say, relation (16), and $\bar{\omega}$ is hyperbolic.

If $\bar{\omega}^2$ is elliptic, X, Y are on $\mathcal{M}$ (**97**). This is also the case when $\bar{\omega}^2$ is a Möbius involution, for (**101**) L_2, M'_2 coincide on $\mathcal{M}$ and the double points X, Y are located on the interior bisector of the angle formed by the lines which join L_2 to two conjugate points L, M'. The symmetry with respect to $\mathcal{M}$ gives

$$(XYL\infty) = \overline{(XYM'\infty)} = \overline{(YX\infty M')},$$

that is to say, equation (17), and $\bar{\omega}$ is elliptic.

Corollary. *$\bar{\omega}$ is parabolic, hyperbolic, or elliptic according as Δ of expression* (12) *or* (13) *is zero, positive, or negative.*

If $\Delta = 0$, we have $\lambda = 1$ and $\bar{\omega}^2$ is parabolic (**94**).

If $\Delta > 0$, λ is real and $\bar{\omega}^2$ is non-involutoric hyperbolic because $\lambda = -1$ implies $a\bar{a} + b\bar{c} + c\bar{b} + d\bar{d} = 0$ and (13) gives

$$\Delta = -4(ad - bc)(\bar{a}\bar{d} - \bar{b}\bar{c}) < 0.$$

If $\Delta < 0$, λ is -1 or $e^{i\theta} \neq 1$ and $\bar{\omega}^2$ is elliptic or involutoric.

124. Elliptic antigraphy. The limit points L, M' may be coincident or distinct (**123**).

Theorem I. *If* $L \equiv M'$, *the antigraphy is the product of an inversion and a rotation having common center* L.

For the origin of axes at L, equation (4) of article **121** is

$$z' = \frac{r}{\bar{z}} e^{i\theta},$$

and we must assume $e^{i\theta} \neq \pm 1$, otherwise we have an inversion (**19**).

Theorem II. *If* $L \neq M'$, *the involutoric pair* PQ *is on the perpendicular bisector of segment* LM' (**123**).

For any two corresponding points Z, Z', *we have the following relations between segments or angles:*

$$\left|\frac{PZ}{QZ}\right| \cdot \left|\frac{PZ'}{QZ'}\right| = \left|\frac{PL}{QL}\right|,$$

$$(ZP, ZQ) - (Z'P, Z'Q) = (LP, LQ).$$

These follow (**25**) from the equation

$$(PQLZ) = \overline{(QP\infty Z')} \quad \text{or} \quad (PQLZ) = \overline{(PQZ'\infty)}.$$

125. Hyperbolic antigraphy. *The double points* E, F *are symmetric with respect to the perpendicular bisector of the segment* LM′ *of the limit points* (**123**).

For any two corresponding points Z, Z′, *we have the following relations between segments or angles:*

$$\left|\frac{EZ}{FZ}\right| : \left|\frac{EZ'}{FZ'}\right| = \left|\frac{EL}{FL}\right|,$$

$$(ZE, ZF) + (Z'E, Z'F) = (LE, LF).$$

These follow from the equation

$$(EFLZ) = (\overline{EF\infty Z'}) \quad \text{or} \quad (EFLZ) = (\overline{FEZ'\infty}).$$

126. Symmetric points. The symmetry with respect to a line and the inversion are the only two involutoric antigraphies. This is why we also give the name symmetry to the inversion, and in place of saying that two points are corresponding or conjugate in an inversion of positive power, we also say that these *points* are *symmetric with respect to the circle* of inversion.

1° *A necessary and sufficient condition for two anharmonic ratios* (ABCP), (ABCQ) *having the same first three points to have conjugate complex numbers for values is that the points* P, Q *be symmetric with respect to the circle or the line which passes through the points* A, B, C.

The condition is sufficient. In fact, if A, B, C are on a circle, the circle is the locus of the double points of an inversion in which A, B, C, P correspond to A, B, C, Q; if A, B, C are on a line, the line is the axis of a symmetry in which A, B, C, P correspond to A, B, C, Q, and in both cases we have (**114**)

$$(ABCP) = (\overline{ABCQ}).$$

The condition is necessary, for if we have the preceding equality and if Q_1 is the symmetric of P with respect to the circle or the line ABC, we also have

$$(ABCP) = (\overline{ABCQ_1}),$$

whence

$$(\overline{ABCQ}) = (\overline{ABCQ_1}), \qquad (ABCQ) = (ABCQ_1)$$

and, consequently (**29**), $Q \equiv Q_1$.

Corollary. *The isodynamic centers* W, W_1 *of a triangle* ABC *are symmetric with respect to the circumcircle.*

In fact we have (**32**)

$$(ABCW) = e^{i\pi/3}, \qquad (ABCW_1) = e^{-i\pi/3}.$$

2° *If two points* P, Q *are symmetric with respect to a line or a circle* γ, *and if we transform the figure by a homography or by an antigraphy, the transforms* P′, Q′ *of* P, Q *are symmetric with respect to the transform* γ' *of* γ.

If A, B, C are three points of γ, we have, by 1°,

$$(ABCP) = \overline{(ABCQ)}. \tag{18}$$

Let A′, B′, C′ be the transforms of A, B, C.

When it is a question of a homography, we have

$$(ABCP) = (A'B'C'P'), \qquad (ABCQ) = (A'B'C'Q')$$

and, by virtue of equation (18),

$$(A'B'C'P') = \overline{(A'B'C'Q')},$$

whence the property, because of 1°.

In the case of an antigraphy,

$$(ABCP) = \overline{(A'B'C'P')}, \qquad (ABCQ) = \overline{(A'B'C'Q')}$$

and, by virtue of (18),

$$\overline{(A'B'C'P')} = (A'B'C'Q').$$

Corollary. *A homography or an antigraphy transforms the vertices* A, B, C *of a triangle into the vertices* A′, B′, C′ *of an equilateral triangle if one of the limit points of the transformation coincides with one of the isodynamic centers* W, W_1 *of* ABC.

In fact, if the homography or the antigraphy is arbitrary, the transforms of W, W_1 are the isodynamic centers W′, W_1' of A′B′C′, for (A′B′C′W′) and (A′B′C′W_1') have the values $e^{\pm i\pi/3}$, and A′B′C′ is equilateral if W′ or W_1' is at infinity (**32**, 2°).

In particular, an inversion of center W or W_1 makes A′B′C′ equilateral, for, since the inverse of the center is the point at infinity, this center is the limit point of the inversion.

127. Determination of the affix of the center of a circle by the method of H. Pflieger-Haertel. Let us be given the circle with parametric equation

$$z = \frac{at + b}{ct + d}, \quad t \text{ real.} \tag{19}$$

The affix ω of its center Ω has already been calculated in (**42**) and by considering Ω as a singular focus of the circle (**61**). A third process consists in considering the circle with equation (19) as the transform of the Ox axis of equation

$$z' = t$$

by the homography with equation

$$z = \frac{az' + b}{cz' + d}. \tag{20}$$

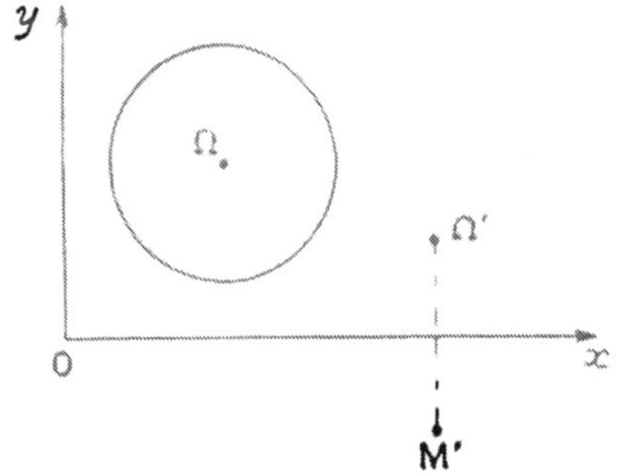

FIG. 59

The center Ω and the point M at infinity in the plane are symmetric with respect to the circle (**126**). Their homologues Ω', M' in the homography (20) are then symmetric with respect to the Ox axis, the homologue of the circle. But the affix of M' being $-d/c$, since it corresponds to z infinite, that of Ω' is $-\bar{d}/\bar{c}$. The affix of Ω is then obtained by replacing z' by $-\bar{d}/\bar{c}$ in (20), and is

$$\omega = \frac{a\bar{d} - b\bar{c}}{c\bar{d} - d\bar{c}}.$$

128. Schick's theorem.[1] *If* A_1, B_1, C_1 *are the orthogonal projections of any point* P *on the sides* BC, CA, AB *of a triangle* ABC, *we have the equality of anharmonic ratios*

$$(ABCP) = \overline{(A_1B_1C_1\infty)}. \tag{21}$$

Let O be the center of the circle (O) circumscribed about triangle ABC and let P' and A_1', B_1', C_1' be the symmetrics of P with respect to (O) and the sides BC, CA, AB.

The first two symmetries give

$$(BCP\infty) = \overline{(BCP'O)} = \overline{(BCA_1\infty)},$$

whence

$$(BCP'O) = (CB\infty A_1')$$

[1] *Münchener Berichte*, 30 (1900), p. 249.

and we consequently have (**100**) the Möbius involution

$$I_a = (BC, OA'_1, P'\infty),$$

of which the two analogous ones are

$$I_b = (CA, OB'_1, P'\infty), \qquad I_c = (AB, OC'_1, P'\infty).$$

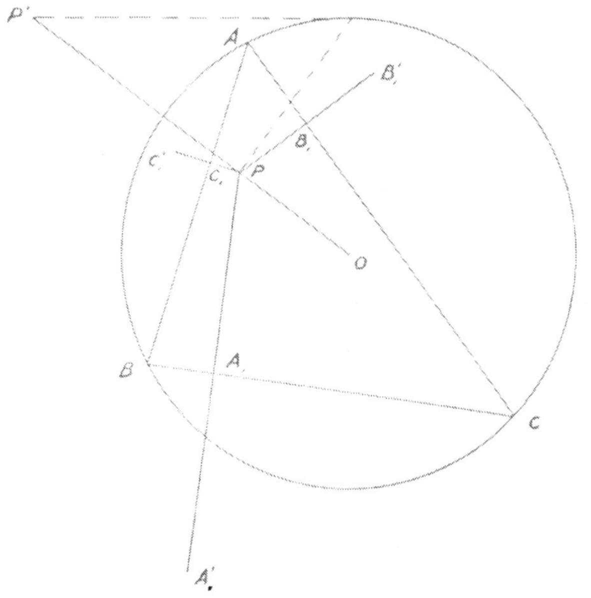

Fig. 60

The product of the involutions I_a, I_b is the homography

$$I_b I_a = \begin{pmatrix} B & A'_1 & P' & \infty \\ A & B'_1 & P' & \infty \end{pmatrix}$$

which gives

$$(P'\infty BA'_1) = (P'\infty AB'_1) = (\infty P'B'_1A),$$

from which we obtain (**100**) the Möbius involution

$$I = (P'\infty, AA'_1, BB'_1).$$

From the homography

$$I_c I_b = \begin{pmatrix} C & B'_1 & P' & \infty \\ B & C'_1 & P' & \infty \end{pmatrix}$$

we obtain in the same way the involution

$$P'\infty, BB'_1, CC'_1$$

which, having two pairs in common with the preceding involution, does not differ from it (**102**), and we therefore have

$$I = (P'\infty,\ AA'_1,\ BB'_1,\ CC'_1).$$

From this we find

$$(ABCP') = (A'_1B'_1C'_1\infty).$$

The inversion in (O) and the homothety with center P and coefficient 1/2 give

$$(ABCP') = (\overline{ABCP}), \qquad (A'_1B'_1C'_1\infty) = (A_1B_1B_1\infty),$$

and hence equation (21) is established.

Corollaries. 1° *The pedal triangle* $A_1B_1C_1$ *of a point* P *is directly similar to the triangle formed by the transforms* A_2, B_2, C_2 *of the vertices of the fundamental triangle in any inversion with center* P.

In fact, the inversion gives

$$(ABCP) = (\overline{A_2B_2C_2\infty})$$

and we therefore have

$$(A_1B_1C_1\infty) = (A_2B_2C_2\infty),$$

which establishes the property (**82**).

2° *If* P', A'_1, B'_1, C'_1 *are the symmetrics of a point* P *with respect to the circumcircle* (O) *of* ABC *and with respect to the sides* BC, CA, AB, *we have the Möbius involution* [1]

$$I = (P'\infty,\ AA'_1,\ BB'_1,\ CC'_1).$$

This involution also possesses the pair PQ, *where* Q *is the second focus of the conic inscribed in triangle* ABC *and whose other focus is* P. [2]

In fact, Q is the center of circle $A'_1B'_1C'_1$. Since this circle is transformed by the involution into the circle ABC, in which P, P′ are inverses, the conjugate of P in I is Q (**126**).

[1] P. Delens, *Mathesis*, 1937, p. 269.
[2] R. Deaux, *Mathesis*, 1954, p. 132.

Exercises 105 through 112

105. Consider the antigraphy $\bar{\omega}$ having distinct limit points L, M′ and equation

$$(\bar{z} - \bar{l})(z' - m') = re^{i\theta}, \qquad r > 0.$$

1° The angle $(\overline{LZ},\overline{M'Z'})$ has the constant value θ.

2° If $\theta \neq k\pi$, pairs of corresponding lines intersect on a circle Γ which passes through L and M′; if $\theta = k\pi$, pairs of corresponding lines are parallel or coincident with LM′.

3° For the origin of axes at the midpoint O of LM′ and Ox on LM′, the parametric equation of Γ is (article 43)

$$z = \frac{l(te^{i\theta} + 1)}{te^{i\theta} - 1};$$

the radius and the affix of the center are $|\, l \csc \theta \,|$ and $-il \cot \theta$.

4° The cartesian equations of $\bar{\omega}$ are

$$x = l + \frac{r[(x' + l) \cos \theta + y' \sin \theta]}{(x' + l)^2 + y'^2},$$

$$y = \frac{-(x' + l) \sin \theta + y' \cos \theta}{(x' + l)^2 + y'^2}.$$

Deduce 3° from this in order to compare the method of classical analytic geometry with the process based on complex numbers.

106. If an antigraphy $\bar{\omega}$ is such that any two of its corresponding lines are parallel (exercise 105), then it is the product of an inversion centered at the limit point L, the power being p, and a translation of vector $\overline{LM'}$, where M′ is the second limit point.
It is parabolic with double point at the midpoint of LM′ if $p = -LM'^2/4$; hyperbolic with double points on LM′ and symmetric with respect to this midpoint if $p > -LM'^2/4$; elliptic with involutoric pair P, Q on the perpendicular bisector of LM′ and having the same midpoint as LM′ if $p < -LM'^2/4$.

107. If an antigraphy has an equation all of whose coefficients are real, it is the product of an inversion centered at a limit point and a translation parallel to the line joining the limit points.
The product of an arbitrary inversion and an arbitray translation has an equation which can be reduced to this form. The same is true of the product of a translation and an inversion.

108. An inversion and a translation are never permutable.

109. The product of an inversion of center I and power p with a rotation of center R and angle θ is an antigraphy which is elliptic if I and R coincide or if

$$p < -IR^2 \qquad \text{or} \qquad p > IR^2 \tan^2 \frac{\theta}{2};$$

hyperbolic if

$$-\mathrm{IR}^2 < p < \mathrm{IR}^2 \tan^2 \frac{\theta}{2};$$

parabolic if

$$p = -\mathrm{IR}^2 \qquad \text{or} \qquad p = \mathrm{IR}^2 \tan^2 \frac{\theta}{2}.$$

Consider the case $\theta = \pi$, and show that the product of the inversion (I,p) and the symmetry of center R is an elliptic, a parabolic, or a hyperbolic antigraphy according as p is less than, equal to, or greater than $-\mathrm{IR}^2$.

110. The product of a rotation of center C and the symmetry with respect to a line d is an involutoric or parabolic antisimilitude according as d does or does not pass through C.

111. 1° If Z, Z′ are any two corresponding points of an antigraphy $\bar{\omega}$ having distinct limit points L, M′, and if L_2, M'_2 are the limit points of $\bar{\omega}^2$, triangles LZL_2, $M'LZ'$ are inversely similar, which yields the construction of L_2 if we know L, M′ and a pair of corresponding points Z, Z′. From this we obtain M'_2 (article **123**).

2° If $\bar{\omega}$ is not the product of an inversion and a translation (exercise 106), the circle Γ on which pairs of corresponding lines of $\bar{\omega}$ intersect (exercise 105) cuts the line $L_2M'_2$ at the double points E, F of $\bar{\omega}$ if $\bar{\omega}$ is hyperbolic or parabolic; the lines LL_2 and $M'M'_2$ are tangent to circle Γ at L and M′; if $\bar{\omega}$ is elliptic, its involutoric pair P, Q is cut on the perpendicular bisector of segment LM′ by the circle with center L_2 and orthogonal to circle Γ.

112. Simson line. 1° A necessary and sufficient condition for the orthogonal projections A_1, B_1, C_1 of a point P on the sides BC, CA, AB of a triangle to be collinear on a line d is that P lie on the circumcircle (O) of ABC [articles **128** and **28**]. Line d is called the *Simson line* of P for ABC.

2° Point A_1 is the midpoint of B_1C_1 if P is the intersection of (O) with the symmedian through A.

3° If (O) is the unit circle, then, with the notation of exercise 41, the equation of d is

$$pz - s_3\bar{z} = \frac{p^3 + s_1p^2 - s_2p - s_3}{2p}.$$

4° The Simson line of the point P′ diametrically opposite P on (O) is perpendicular to d and cuts d on the Feuerbach circle (exercise 43) at the point with affix

$$\omega = \frac{s_1 - \dfrac{s_3}{p^2}}{2}.$$

The equation of d can be written as

$$\frac{z - \omega}{\bar{z} - \bar{\omega}} = \frac{s_3}{p}.$$

VII. *PRODUCT OF SYMMETRIES*

129. Symmetries with respect to two lines. **I.** *The product of two symmetries with respect to parallel axes* d_1, d_2 *is the translation whose vector is twice that of the translation perpendicular to* d_1 *and which carries* d_1 *into* d_2.

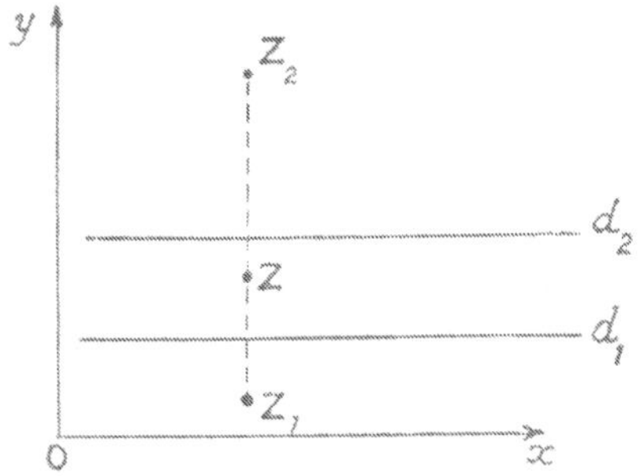

FIG. 61

Take Ox parallel to d_1, d_2 and let a_1, a_2 be the ordinates of these lines, whose equations are then (**33**)

$$z - \bar{z} = 2ia_1,$$
$$z - \bar{z} = 2ia_2.$$

If Z_1 is the symmetric of any point Z with respect to d_1, and Z_2 is the symmetric of Z_1 with respect to d_2, the equations of the two symmetries are (**117**)

$$z_1 - \bar{z} = 2ia_1, \qquad z_2 - \bar{z}_1 = 2ia_2,$$

whence the equation of their product, taken in the indicated order, is

$$z_2 = z + 2i(a_2 - a_1),$$

which proves the theorem (**14**).

Conversely, *a given translation is, in an infinity of ways, the product of two symmetries with respect to two axes perpendicular to the direction of the translation; the first axis being chosen arbitrarily, the second is obtained by a translation whose vector is half that of the given translation.*

II. *The product of two symmetries with respect to axes* d_1, d_2 *which intersect in a proper point* O *is the rotation with center* O *and with angle twice the algebraic value* (d_1d_2) *of the angle formed by the arbitrarily oriented axes* d_1, d_2.

Take the origin of the coordinate axes at O and let α_1, α_2 be the angles (xd_1), (xd_2). Line d_1, passing through the origin, has an equation of the form (**37**)

$$az + \bar{a}\bar{z} = 0$$

and is therefore

$$\frac{z}{\bar{z}} = \frac{x+iy}{x-iy} = \frac{OA_1e^{i\alpha_1}}{OA_1e^{-i\alpha_1}} = e^{2i\alpha_1}.$$

Using the same notation employed in the preceding theorem, the two symmetries then have for equations

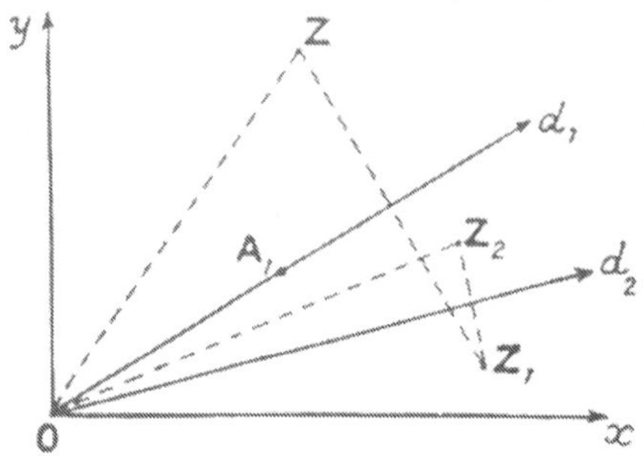

FIG. 62

$$z_1 = \bar{z}e^{2i\alpha_1}, \quad z_2 = \bar{z}_1 e^{2i\alpha_2}$$

and the equation of their product is

$$z_2 = ze^{2i(\alpha_2-\alpha_1)},$$

whence the theorem (15).

Conversely, *a given rotation is, in an infinity of ways, the product of two symmetries with respect to axes issuing from the center of the rotation; the first axis being chosen arbitrarily, the second is obtained by a rotation whose angle is half that of the given rotation.*

130. Symmetry and inversion. *The product of the symmetry with respect to a line* d *and the inversion of power* p *and center* O *not on* d *is a hyperbolic, elliptic, or parabolic homography according as* p *is less than, greater than, or equal to the square of the distance of* O *from* d.

If O *is on* d, *the product is a Möbius involution having* O *for central point, and whose double points are on* d *or on the perpendicular to* d *through* O *according as the power* p *is positive or negative.*

We have similar results for the product of an inversion and a symmetry with respect to a line.

Take Ox parallel to d and let a be the ordinate of a point of d. The equations of the symmetry and the inversion being (**129, 19**)

$$z_1 - \bar{z} = 2ia,$$
$$z'\bar{z}_1 = p,$$

the equation of the product is

$$z'(z - 2ia) = p.$$

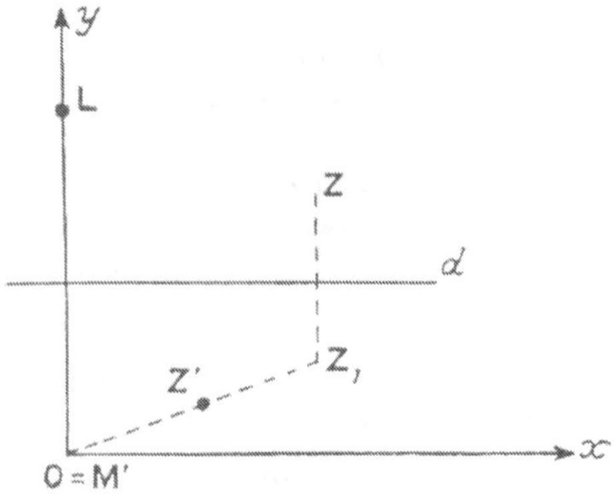

FIG. 63

We thus have a homography whose limit points L, M′ have $2ia$, 0 for affixes, and whose double points E, F have the roots of

$$z^2 - 2iaz - p = 0$$

for affixes, that is to say,

$$ia \pm \sqrt{p - a^2}.$$

If a is not zero, that is, if the center O is not on the axis d, the homography is not involutoric; points E, F are distinct on Oy or d according as p is less than or greater than a^2; E, F coincide at the point (d, Oy) if $p = a^2$.

If $a = 0$, we have a Möbius involution with equation

$$z'z = p.$$

The central point is O and points E, F, with affixes $\pm \sqrt{p}$, are on Ox or Oy according as p is a positive or a negative number.

Conversely, *a non-loxodromic and non-similitude homography is, in a unique way, the product of a symmetry with respect to a line and an inversion with center* M′, *and, in a unique way, the product of an inversion with center* L *and a symmetry with respect to a line* (**95, 96, 97**). *The symmetry and the inversion are permutable if the homography is a Möbius involution* (**101**).

131. Product of two inversions. An inversion with center O and negative power p has no real double points, for its equation

$$z'\bar{z} = p$$

for origin at O gives the double points by means of the equation

$$z\bar{z} = p \quad \text{or} \quad x^2 + y^2 = p \qquad (1)$$

which has no real roots. Since equation (1) has real coefficients, it represents an *ideal circle* (**40**), whose center O is real, the square of whose radius is the negative number p, and all of whose points are imaginary and are said to be double for the inversion. This ideal circle is called the circle of inversion. By considering such circles of inversion, we can employ properties about pencils of circles established in analytic geometry or in projective geometry when these pencils do not exclude the imaginaries of von Staudt.

Theorem. *The product* I_2I_1 (**21**) *of two inversions* I_1, I_2 *having the distinct points* O_1, O_2 *as centers,*[1] p_1, p_2 *for powers, and* γ_1, γ_2 *for circles of inversion is a non-loxodromic homography* ω *having for limit points* L *and* M′ *the inverses, respectively, of* O_2 *and* O_1 *in* I_1 *and* I_2, *and for double points* E *and* F *the centers of the null circles (the Poncelet points) or the base points of the pencil of circles to which* γ_1 *and* γ_2 *belong.*

[1] If $O_1 = O_2$, the product I_2I_1 is the homothety (O_1, p_2/p_1).

The homography ω *is*

1° *hyperbolic if at least one of the circles* γ_1, γ_2 *is ideal or if* γ_1, γ_2 *are real and non-intersecting;*

2° *parabolic if* γ_1, γ_2 *are real and tangent;*

3° *elliptic if* γ_1, γ_2 *are real and intersecting;*

4° *a Möbius involution if the circles* γ_1, γ_2 *are orthogonal, one of the circles being allowed to be ideal.*

If L_1 is the inverse of L in I_1, the inverse of L_1 in I_2 must be at infinity. Hence $L_1 = O_2$, and

$$O_1L.O_1O_2 = p_1. \tag{2}$$

Since O_1 is the inverse, in I_1, of the point at infinity, M′ is the inverse of O_1 in I_2, and we have

$$O_2M'.O_2O_1 = p_2. \tag{3}$$

The homography ω is a Möbius involution if $L = M'$. Equation (3) then gives

$$LO_2.O_1O_2 = p_2$$

and, by adding this to (2), we find

$$p_1 + p_2 = O_1O_2^2. \tag{4}$$

Since p_1, p_2 are the squares of the radii of γ_1, γ_2, these circles are orthogonal.

Line O_1O_2 contains the homologue in ω of each of its points, and is thus a double line for ω. Hence ω is non-loxodromic. The double points E, F of ω are therefore on line O_1O_2 or on the perpendicular to O_1O_2 at the midpoint of segment LM′, a fact which will determine the hyperbolic, parabolic, or elliptic character of ω.

If E, F are on O_1O_2, they constitute a pair of conjugate points in each of the inversions, for if $O_1E.O_1E_1 = p_1$ we must also have $O_2E_1.O_2E = p_2$ so that E may be double for ω. It follows that E, F will be on O_1O_2 if there exists on this line two points which are conjugate with respect to each of the circles γ_1, γ_2, in other words, if the pencil of circles determined by γ_1, γ_2 contains two point circles, distinct or not; these point circles will be the points E, F. This will occur when at least one of the circles γ_1, γ_2 is ideal, or when γ_1, γ_2 are real but non-intersecting, or when γ_1, γ_2 are tangent, in which case E = F.

If γ_1, γ_2 are real and intersecting, their common points are necessarily E and F.

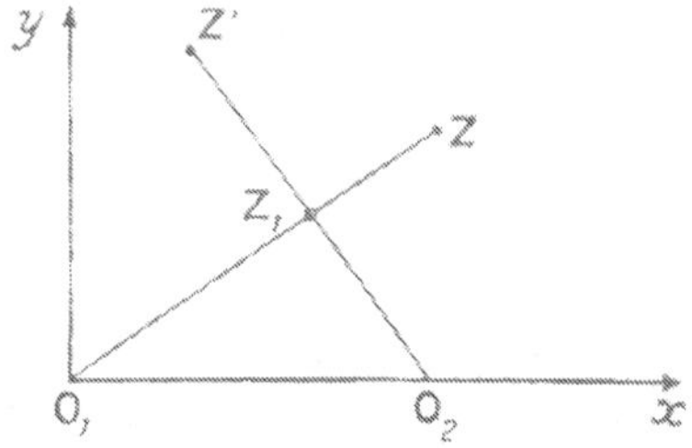

FIG. 64

These results can also be obtained by analytical methods. Take O_1 for origin, place O_1x on line O_1O_2 and let $O_1O_2 = a$. The equations of I_1, I_2 being

$$z\bar{z}_1 = p_1, \qquad (z' - a)(\bar{z}_1 - a) = p_2,$$

that of ω is

$$azz' - (a^2 - p_2)z - p_1 z' + ap_1 = 0 \tag{5}$$

and the affixes of E, F, the roots of

$$az^2 - (a^2 - p_2 + p_1)z + ap_1 = 0, \tag{6}$$

are real and distinct, real and equal, or conjugate imaginary according as

$$\Delta = (a^2 - p_2 + p_1)^2 - 4a^2p_1 = (p_1 + p_2 - a^2)^2 - 4p_1p_2 =$$
$$(p_1 - p_2)^2 + a^4 - 2a^2(p_1 + p_2) \tag{7}$$

is positive, zero, or negative.

If at least one of the numbers p_1, p_2 is negative, Δ is obviously positive. If p_1, p_2 are positive and equal to the squares r_1^2, r_2^2 of the radii of the real circles γ_1, γ_2, we have

$$\Delta = (r_1^2 + r_2^2 - a^2)^2 - 4\,r_1^2r_2^2 =$$
$$(r_1 + r_2 + a)(r_1 + r_2 - a)(r_1 - r_2 + a)(r_1 - r_2 - a)$$

positive, zero, or negative according as γ_1, γ_2 do not intersect, are tangent, or do intersect.

Corollaries. 1° *If the product* I_2I_1 *is a Möbius involution, the double points* E, F *of the involution are on the line* O_1O_2 *if one of the inversions has a negative power, and on the perpendicular to* O_1O_2 *at* L = M′ *if the powers* p_1, p_2 *are both positive. We have*

$$(O_1O_2EF) = -1.$$

Because of (4), expression (7) for Δ becomes $-4p_1p_2$; E, F are then on O_1O_2 if $p_1p_2 < 0$.

Since O_2 is the homologue of O_1 in ω, we have $(EFO_1O_2) = -$.

2° *A necessary and sufficient condition for the product of the inversions* I_1, I_2 *to be a Möbius involution is that the inversions be permutable.* In fact, the product I_2I_1 is an involutoric homography if (**23**)

$$(I_2I_1)^2 = 1 \quad \text{or} \quad I_2I_1I_2I_1 = 1.$$

If we multiply on the right by I_1, then on the left by I_2, and note that $I_1^2 = 1 = I_2$, we have

$$I_1I_2 = I_2I_1,$$

and conversely.

132. Homography obtained as product of inversions. **Theorem I.** *A hyperbolic, parabolic, or elliptic homography is, in a single infinity of ways, the product of two inversions whose centers are found on the line joining the limit points of the homography.*

Take any point O_1 on the line joining the limit points L, M′ of the homography ω whose double points are E, F. Let I_1 be the inversion with center O_1 and power $p_1 = O_1E.O_1F$ or O_1E^2 according as ω is or is not hyperbolic. The inversion I_2 having for center the point O_2 such that $O_1O_2.O_1L = p_1$ and for power $p_2 = O_2E.O_2F$ or O_2E^2 according as ω is or is not hyperbolic, is such that

$$I_2I_1 = \omega.$$

In fact, the homography I_2I_1 has E, F for double points and L for limit point (**131**), and is therefore ω (**74**).

Theorem II. *There are two single infinities of ways of obtaining a Möbius involution as a product of two inversions. Either the centers are on the line determined by the double points* E, F, *which they separate harmonically, and each of the inversions interchanges* E *and* F, *or the circles of inversion intersect orthogonally at* E *and* F (**131**, corollary 1°).

Theorem III. *Every loxodromic homography is, in an infinity of ways, the product of four inversions.*

It is, in fact, in a double infinity of ways, the product of two Möbius involutions (**110**, corollary), and each of these involutions is obtained as a product of two inversions (theorem II).

133. Antigraphy obtained as product of three symmetries. **Theorem I.** *Every antigraphy* $\bar{\omega}$ *is the product of three symmetries.*

1° Suppose, first of all, that $\bar{\omega}$ is not an antisimilitude; it then does not have the point at infinity for a double point (**116**). If

S designates the symmetry with respect to an arbitrarily chosen line s, we have

$$\bar{\omega} = \omega S,$$

ω being a non-similitude homography since otherwise the point at infinity would be a double point for $\bar{\omega}$. We know (**110**) that ω is, in an infinity of ways, the product J_2J_1 of two Möbius involutions which have the double points of ω for pairs of conjugate points, one of them, say J_1, being able to be determined by the arbitrary choice of a second pair of conjugate points. Since ω is not a similitude, we can choose J_1 so that its central point will be on s. We have

$$\bar{\omega} = J_2J_1S.$$

But J_1S is an inversion (**101**) I_1, whence

$$\bar{\omega} = J_2I_1.$$

Since J_2 is (**132, II**), in an infinity of ways, the product I_3I_2 of two inversions, we have

$$\bar{\omega} = I_3I_2I_1.$$

2° If $\bar{\omega}$ is an antisimilitude, its transform (**112**) by a non-similitude homography θ is a non-antisimilitude antigraphy $\bar{\omega}'$, the product $I_3'I_2'I_1'$ of three symmetries whose transforms by θ^{-1} give $\bar{\omega}$ as the product of three symmetries.

Corollary. *A homography is the product of four symmetries (axial symmetries or inversions)*, for it is the product of an axial symmetry and an antigraphy.

Theorem II. *The product* $I_3I_2I_1$ *of three inversions having distinct centers is an inversion* I_4 *or a symmetry* I_4 *with respect to a line if the circles of inversion* (I_1), (I_2), (I_3) *belong to a common pencil of circles or are orthogonal in pairs. The circle or the line* (I_4) *belongs to the same pencil or is orthogonal to* (I_1), (I_2), (I_3).

The equation

$$I_3I_2I_1 = I_4$$

is equivalent to

$$I_3I_2 = I_4I_1$$

obtained by multiplying the two sides of the preceding equation on the right by I_1. The homography $\omega = I_3I_2$ is neither loxodromic nor a similitude (**131**).

If ω is non-involutoric and hyperbolic, or is parabolic, the circles (I_2), (I_3) define a pencil of circles having the double points E, F of ω for Poncelet points (**131**), and since we must have $\omega = I_4 I_1$, it is necessary that the circles (I_1), (I_4) belong to this pencil. Moreover, it is sufficient that (I_1) belong to this pencil in order that I_4 exist (**132**).

If ω is elliptic, a similar argument proves that (I_1), (I_2), (I_3) must pass through E, F, a condition which is sufficient.

If ω is a Möbius involution, it is possible for the orthogonal circles (I_2), (I_3) to have E, F for Poncelet points or to pass through E, F, the first situation holding if (I_2) or (I_3) is ideal (**131**). Similarly, the orthogonal circles (I_1), (I_4) belong to the same pencil as do (I_2), (I_3) or are orthogonal to these circles (**132, II**).

Assorted exercises 113 through 136

113. The product of an inversion and a translation is the only antigraphy (elliptic, hyperbolic, or parabolic) whose limit points have the same midpoint as the involutoric pair or as the double points.

114. If A, B, C are the vertices of a triangle, the antigraphy

$$\bar{\omega} = \begin{pmatrix} A & B & C \\ B & C & A \end{pmatrix}$$

is elliptic. The points of the involutoric pair are the isodynamic centers and the limit points are the Brocard points. The point of intersection of a pair of corresponding lines describes the Brocard circle. The square of $\bar{\omega}$ is a cyclic homography of period 3.

115. Boutin points. There exist, on the circumcircle (O) of a triangle ABC, three points B_i, called *Boutin points*, such that OB_i is parallel to the Simson line of B_i. $B_1B_2B_3$ is an equilateral triangle. If we take (O) as unit circle and a point B_i as unit point (exercise 40), then $s_3 = 1$, for if the unit point is arbitrary on (O) the affixes of B_i are roots of $z^3 = s_3$.

116. The tangent to the unit circle (O) at the point M with affix e^{it} has the equation

$$z + \bar{z}e^{2it} = 2e^{it}.$$

Let M_1 be the orthogonal projection of M on the diameter Ox. Show that M_2, the symmetric of M_1 with respect to the considered tangent, describes a *two-cusped epicycloid* $\mathscr{E}$. Determine the base curve and the generating curve. Show that the tangent to (O) at M and the tangent to $\mathscr{E}$ at M_2 intersect on Ox.

117. Two vectors $\overline{OA}$, $\overline{OB}$ of the same length rotate about O with constant angular velocities of algebraic values α, β such that $\alpha + \beta \neq 0$. Take the circle with center O and radius OA as unit circle. Let a, b be the affixes of A, B and set $e^{it} = \tau$, t being a real parameter.

1° Line AB has the equation

$$z + ab\tau^{\alpha+\beta}\bar{z} = a\tau^{\alpha} + b\tau^{\beta}.$$

2° This line envelops a cycloidal curve Γ with equation

$$z = \frac{a\beta\tau^{\alpha} + b\alpha\tau^{\beta}}{\alpha + \beta}$$

and its point of contact T divides segment AB in the constant ratio $\alpha : \beta$. Curve Γ is tangent to the circle at the points given by

$$\tau^{\beta-\alpha} = \frac{a}{b};$$

if one of these points of contact is chosen for unit point, the equation of Γ can be written as

$$z = \frac{\beta\tau^{\alpha} + \alpha\tau^{\beta}}{\alpha + \beta}.$$

118. Two points P, Q, which are initially coincident at a point A of a circle Γ, move on the circle in the same direction with constant angular velocities, that of Q being twice that of P. Show that the point M situated at the trisection point of segment PQ nearer P describes a cardioid tangent to PQ at M, whose singular focus is the center O of Γ and whose cusp R is on AO such that OR = AO/3.

119. Two points P, Q, starting from a common initial point A on a circle (O) of center O and radius 1, move on the circle in opposite directions, the speed of Q being twice that of P. The symmetric of Q with respect to P describes a *hypocycloid of three cusps* (a *Steiner hypocycloid*, or *deltoid*), tritangent to circle (O) at three points, one of which is A, which are vertices of an equilateral triangle AA′A″. The cusps R, R′, R″ are the homologues of A, A′, A″ in the homothety having center O and coefficient — 3.

If the Ox axis is placed on OA from O toward A, then the equation of the hypocycloid is

$$z = 2e^{it} - e^{-2it}.$$

Show, by changing the parameter, that this curve is a unicursal circular quartic of third class with no foci.

The orthogonal projection of O on the line PQ describes a *regular trifolium* (or *three-leaved rose*) having equation

$$z = \frac{e^{it} + e^{-2it}}{2},$$

and is a unicursal circular quartic of the sixth class having a triple point at O with the consecutive tangents making angles of 120° with one another. It can be generated as follows : a circle ρ_1 with center A_1 and radius 1/4 rolls without sliding on the fixed circle β_1 with center O and the same radius; the symmetric of O with respect to A_1 and invariably fixed to ρ_1 generates the trifolium.

120. Let us take the circumcircle of a triangle ABC as unit circle and the point diametrically opposite a Boutin point (exercise 115) as unit point, and let us denote the affix of a point P of the circle by e^{it}. Then the envelope of the Simson line of P is a Steiner hypocycloid having equation

$$z = \frac{s_1}{2} + \frac{2e^{it} - e^{-2it}}{2}.$$

It is tangent to the sides of the triangle and tritangent to the Feuerbach circle of center O_9 at the points A_i such that, if B_i are the Boutin points, $\overline{O_9A_i} = \overline{B_iO}/2$.

121. If the circumcircle (O) of a triangle ABC with centroid G is the unit circle, and if the Ox axis contains a Boutin point B_1, then the number $s_2/3$ is the affix of the symmetric of G with respect to OB_1.

122. 1° The circumcircle (O) of a triangle ABC being the unit circle, the affix of the centroid G_1 of the *pedal triangle* of any point P (formed by the orthogonal projections A_1, B_1, C_1 of P on BC, CA, AB) is

$$\frac{s_1}{3} + \tfrac{1}{2}(p - s_2\bar{p}/3).$$

2° If P describes (O), of radius R, the centroid of the collinear points A_1, B_1, C_1 moves on an ellipse $\mathscr{E}$ centered at the centroid G of triangle ABC. The axes of $\mathscr{E}$ have lengths R ± OG and are parallel to the bisectors of the angles formed by a Boutin diameter OB_i and the symmetric of OG with respect to this diameter. $\mathscr{E}$ passes through the trisection points of the altitudes of ABC which are nearer the vertices, through the orthogonal projections on BC, CA, AB of the points of intersection of the symmedians with (O), and through the symmetrics, with respect to the midpoints of the sides, of the points which divide the distances from these midpoints to the feet of the corresponding altitudes in the ratio 1 : 2.

123. 1° The antigraphy $\bar{\omega}$ which associates with the vertices A, B, C of a triangle the vertices A′, B′, C′ of the pedal triangle of a point P has the equation

$$2z'\bar{z} - 2\bar{p}z' - (s_1 - s_2\bar{p} + s_3\bar{p}^2 + p)\bar{z} + 1 + p\bar{p} = 0.$$

The affixes of the limit points L, M′ are

$$l = p, \qquad m' = \frac{s_1 - s_2\bar{p} + s_3\bar{p}^2 + p}{2}.$$

2° $\bar{\omega}$ is an inversion only if P is the orthocenter of ABC. Determine the circle of inversion.

3° $\bar{\omega}$ is never the product of an inversion and a concentric rotation.

4° The homography which associates A′, B′, C′ with A, B, C has the equation

$$2\bar{p}z''z - 2z'' - (1 + p\bar{p})z + s_1 - s_2\bar{p} + s_3\bar{p}^2 + p = 0.$$

It is a Möbius involution only if P is on the circumcircle of ABC.

124. In a triangle ABC, determine the pairs of points for which the vertices of the pedal triangle can be so associated as to give equal triangles. [See *Mathesis*, 1949, p. 257.]

125. If a conic is tangent at A′, B′, C′ to the sides BC, CA, AB of a triangle, its foci constitute a pair of corresponding points in each of the six Möbius involutions

(AA,B′C′), (BB,C′A′), (CC,A′B′), (BC,AA′), (CA,BB′), (AB,CC′).

126. An arbitrary affinity which transforms the Gauss plane into itself has an equation of the form

$$z' = az + b\bar{z} + c,$$

where a, b, c are complex constants whose images yield a simple description of the properties of the affinity. [See *Mathesis*, 1950, p. 101.]

127. A variable point P in the plane of a triangle ABC and the centroid of its pedal triangle correspond in an affinity. State some of the properties of this transformation. [See exercises 122, 126.]

128. The pairs of lines OA, OA_1 and OB, OB_1 form angles having the same bisectors if

$$\frac{aa_1}{\bar{a}\bar{a}_1} = \frac{bb_1}{\bar{b}\bar{b}_1},$$

where a, a_1, b, b_1 are the affixes of A, A_1, B, B_1 in a rectangular system (Ox, Oy).

129. Isogonal points. Let P be any point in the plane of a triangle ABC and consider the point Q such that the pairs of lines (AP,AQ), (AB,AC) form two angles having the same bisectors, and similarly for the pairs of lines (BP,BQ), (BC,BA).

1° If the circumcircle of triangle ABC is the unit circle (exercise 40), we have (exercise 128)

$$pq - a(p + q) + a^2 = [\bar{p}\bar{q} - \bar{a}(\bar{p} + \bar{q}) + \bar{a}^2]as_3,$$

$$pq - b(p + q) + b^2 = [\bar{p}\bar{q} - \bar{b}(\bar{p} + \bar{q}) + \bar{b}^2]bs_3.$$

Show that

$$p + q + \overline{pq}s_3 = s_1.$$

2° The angles (CP,CQ), (CA,CB) also have the same bisectors. The points P, Q are said to be *isogonal points* in triangle ABC. The transformation which associates with each point P its isogonal Q is called an *isogonal point transformation* (or an *isogonal inversion*). P and Q are the foci of a conic inscribed in the triangle, whence the name *focal pairing* given to the transformation by F. and F. V. MORLEY (*Inversive Geometry*, 1933, p. 196). It has equation (1).

130. If P is a given focus of a conic Γ inscribed in triangle ABC, the affix of the second focus Q is given by

$$\bar{q} = \frac{s_3(p^2 - s_1 p + s_2 - \bar{p}s_3)}{1 - p\bar{p}}$$

or, if we set

$$p^3 - s_1p^2 + s_2p - s_3 = (p - a)(p - b)(p - c) = \Pi,$$

by

$$\bar{q} = \frac{1}{p} + \frac{\bar{s}_3\Pi}{p(1 - p\bar{p})}.$$

When $|p| = 1$, P is on the circumcircle, Γ is a parabola, and Q is the point at infinity.

131. The locus of pairs of isogonal points collinear with the circumcenter of triangle ABC has the equation (exercise 130)

$$(z - a)(z - b)(z - c) = s_3t,$$

where t is a real parameter. This is a non-circular cubic called the *MacCay cubic* of the triangle. It passes through the vertices A, B, C of the triangle, where it touches the altitudes, through the orthocenter H, through the circumcenter O, where the tangent is OH, through the centers of the inscribed and escribed circles, where the tangents contain O, through the points (BC,OA), (CA,OB), (AB,OC), and through the points on (O) diametrically opposite the Boutin points. The asymptotes are parallel to the Boutin diameters and pass through the centroid of ABC. [See *Mathesis*, 1949, p. 225.]

132. Let $\bar{\omega}$ be the antigraphy which associates with the vertices A, B, C of a triangle the vertices A′, B′, C′ of the pedal triangle of a point P whose isogonal is Q.

1° The affix of the limit point M′ is (exercises 123, 130), if $p \neq 0$,

$$m' = \frac{1 + p\bar{p} + \bar{\Pi}s_3}{2\bar{p}}.$$

2° The vectors $\overline{PM'}$, $\overline{OQ}$ have the same direction and, if R is the radius of the circumcircle (O) of ABC, we have, in algebraic value,

$$PM' = \frac{R^2 - OP^2}{2R^2} OQ.$$

3° The equation giving the double points or the involutoric pair is

$$2(\bar{m}' - \bar{p})z^2 - 2(m'\bar{m}' - p\bar{p})z + (1 + p\bar{p})(m' - p) = 0.$$

4° The locus of points P such that $\bar{\omega}$ may be the product of an inversion and a translation is the MacCay cubic of the triangle. [Use exercises 113 and 131.] P is the center of the inversion and the vector of the translation is

$$\frac{R^2 - OP^2}{2R^2} \overline{OQ}.$$

Lines PA, M′A′ are parallel, as are PB, M′B′ and PC, M′C′.

133. Being given the affix m of the center of a conic inscribed in a triangle ABC, form the equation giving the affixes of the foci.

Find, in this way, the equation giving the foci of the inscribed Steiner ellipse (exercise 84) in the case where the circumcircle is the unit circle.

In the same case, the affixes of the centers of the circles tangent to the sides of the triangle are the roots of the equation

$$2z + s_3\bar{z}^2 = s_1,$$

which is equivalent to

$$z^4 - 2s_2z^2 + 8s_3z + s_2^2 - 4s_1s_3 = 0.$$

134. A triangle ABC and a line which cuts the sides BC, CA, AB in A′, B′, C′ form a quadrilateral Q giving the Möbius involution (AA′, BB′, CC′).

1° The equation of this involution is

$$(z - a)(z - a') = \lambda(z - b)(z - b'),$$

where λ represents a complex parameter, and the affix of the Miquel point M is any one of the expressions

$$\frac{aa' - bb'}{a + a' - b - b'},$$

$$\frac{bb' - cc'}{b + b' - c - c'},$$

$$\frac{cc' - aa'}{c + c' - a - a'}.$$

2° A focus Z of a conic inscribed in Q being such that, for example, angles (ZA,ZB), (ZB′,ZA′) are equal, show (exercise 128) that the locus of the foci of all these conics has the equation

$$(z - a)(z - a') = t(z - b)(z - b'),$$

where t denotes a real parameter.

This locus, called the *focal locus of Van Rees*, is, in general, a non-unicursal circular cubic. It passes through the six vertices of Q and the Miquel point, which is the focus of the parabola of the family of conics.

3° The two (real) foci of any one of the conics are given by the same value of t (article 98).

4° The focal locus of Van Rees is the locus of pairs of conjugate points in a Möbius involution. The midpoints of the pairs of points are collinear.

5° The locus of the harmonic conjugate of a fixed point P with respect to the real foci of a variable conic inscribed in a quadrilateral Q is a straight line if Q is a parallelogram; otherwise it is a straight line or a circle according as P is or is not on the line containing the midpoints of the diagonals of Q.

6° The harmonic conjugates of a point with respect to the three pairs of opposite vertices of a quadrilateral lie on a straight line or a circle which passes through the symmetric of the point with respect to the Miquel point of the quadrilateral. [See *Nieuw tijdschrift voor wiskunde*, 35th year, 1947-48, p. 151.]

135. A necessary and sufficient condition for the two pairs of real foci of two conics inscribed in a triangle to separate one another harmonically is that the centers of the conics be a pair of isogonal points. [See *Mathesis*, 1954, pp. 218, 311.]

136. If C is an arbitrary point in the plane of the Möbius involution I = (AA′,BB′), and if D is the second point of intersection of the two circles AB′C, A′BC, then the second point of intersection of the two circles ABD, A′B′D is the conjugate C′ of C in I (STRUBECKER, *Monatshefte für Math. und Physik*, vol. 41, 1934, p. 439). [It suffices to consider the transform I_1 of I by an inversion of center C, and the central point of I_1 (article 104).]

INDEX

References are to articles or to exercises (E)

introductiontoge00rola
introductiontoge00rola

Lightning Source UK Ltd.
Milton Keynes UK
UKHW022112160223
417167UK00006B/107

9 781014 126436